Ravi S. Kulkarni, Ulrich Pinkall (Eds.)

Conformal Geometry

Aspects of Mathematics

Aspekte der Mathematik

Editor: Klas Diederich

All volumes of the series are listed on pages 237–238.

Ravi S. Kulkarni, Ulrich Pinkall (Eds.)

Conformal Geometry

A Publication of the Max-Planck-Institut für Mathematik, Bonn
Adviser: Friedrich Hirzebruch

Friedr. Vieweg & Sohn Braunschweig / Wiesbaden

CIP-Titelaufnahme der Deutschen Bibliothek

Conformal geometry:
a publ. of the Max-Planck-Inst. für Mathematik,
Bonn / Ravi S. Kulkarni; Ulrich Pinkall (ed.).
Advisor: Friedrich Hirzebruch. – Braunschweig;
Wiesbaden: Vieweg, 1988
(Aspects of mathematics: E; Vol. 12)
ISBN 3-528-08982-2

NE: Kulkarni, Ravi S. [Hrsg.]; Max-Planck-Institut für Mathematik ⟨Bonn⟩; Aspects of mathematics / E

AMS Subject Classification: 53 A 30

Vieweg is a subsidiary company of the Bertelsmann Publishing Group.

Produced by W. Langelüddecke, Braunschweig
Printed in Germany

ISSN 0179-2156

ISBN 3-528-08982-2

Contents

The reader can find a detailed table of contents at the beginning of each article.

Authors of this Volume

Wolfgang Kühnel
Fachbereich Mathematik
Universität Duisburg
D-4100 Duisburg 1
FRG

Ravi S. Kulkarni
Department of Mathematics
City University of New York,
Queens College,
Flushing, NY 11367
USA

Jacques Lafontaine
U.E.R. de Mathématique
Université Paris 7
F-75251 Paris Cedex 05
France

Ulrich Pinkall
Fachbereich Mathematik
TU Berlin
Straße des 17. Juni 135
D-1000 Berlin 12

Hans-Bert Rademacher
Mathematisches Institut
Universität Bonn
Wegelerstr. 10
D-5300 Bonn 1
FRG

Seppo Rickman
Department of Mathematics
University of Helsinki
SF-00100 Helsinki
Finland

Preface

The contributions in this volume summarize parts of a seminar on conformal geometry which was held at the Max-Planck-Institut für Mathematik in Bonn during the academic year 1985/86. The intention of this seminar was to study conformal structures on manifolds from various viewpoints. The motivation to publish seminar notes grew out of the fact that in spite of the basic importance of this field to many topics of current interest (low-dimensional topology, analysis on manifolds ...) there seems to be no coherent introduction to conformal geometry in the literature.

We have tried to make the material presented in this book self-contained, so it should be accessible to students with some background in differential geometry. Moreover, we hope that it will be useful as a reference and as a source of inspiration for further research.

Ravi Kulkarni/Ulrich Pinkall

Conformal Structures and Möbius Structures

*Ravi S. Kulkarni**

Contents

---

* Partially supported by the Max-Planck-Institut für Mathematik, Bonn, and an NSF grant.

§0 Introduction

(0.1) Historically, the stereographic projection and the Mercator projection must have appeared to mathematicians very startling. It was an indication that the conformal maps among the surfaces have far more flexibility than for example the isometries among surfaces, or line-preserving maps among planar domains. This was confirmed by Gauss in his *A general solution to the problem of mapping parts of a given surface onto another surface such that the image and the mapped parts are similar in the smallest parts.* This is esentially the existence of "isothermal co-ordinates" in the C^ω case. It is interesting to note that this study preceded and partially motivated Gauss's later foundational work on the notion of curvature. For an account of this interesting history see Dombrowski [D], pp 127-130.

(0.2) Another equally startling discovery is the connection of isothermal coordinates to an entirely different idea, namely that of a holomorphic function of a complex variable. For example, the Mercator projection is essentially the holomorphic map $z \mapsto \log z$! The global aspect of this is the theory of Riemann surfaces.

(0.3) Still another startling fact is that a compact Riemann surface is essentially the same as a complex projective algebraic curve.

(0.4) These deep local and global connections set up one of the natural goals of conformal geometry: namely, to understand the differential-geometric underpinnings of these classical theories–i.e., to separate the analytic aspects from the topological ones as clearly as possible, and relate them to the more primitive geometric notions of distance, angle, area, straight line ..., and equally importantly, to isolate the role of symmetry. The symmetry considerations here mainly concern the groups of Möbius transformations–in particular, the classical theories of Fuchsian and Kleinian groups. Due to certain isomorphisms such as $M_o(1) \approx PSL_2(R) \approx SO_o(2,1)$, $M_o(2) \approx PSL_2(C) \approx SO_o(3,1) \cdots$ the conformal considerations have turned out to be basic for 3- and 4-dimensional topology, and in the physics of relativity.

(0.5) In the first seven sections of this chapter we have striven to bring together the traditional differential-geometric, and the traditional Kleinian-group theoretic viewpoints. These two traditions have developed (until quite recently) quite independently. Their mutual interaction should prove to be fruitful. We have isolated the notion of a "Möbius structure" from the general conformal structure. A geometer would find it useful to keep in mind the following facts.

i) A 2-dimensional conformal structure is always integrable, but is ambient to several Möbius structures. (Example: the stereographic projection is circle-preserving, i.e., a Möbius map, whereas the Mercator projection preserves angles but not circles, i.e., is not a Möbius map.)

ii) In dimensions ≥ 3, a conformal structure is not always integrable, but an integrable one is ambient to a unique Möbius structure. The latter is essentially the Liouville's theorem. The proof of Liouville's theorem given here is quite elementary and works in the C^2 case.[1] In general, the "Möbius" arguments, provided they work at all, are aften simpler and give better results.

The last section indicates some recent developments in the joint work with U. Pinkall.

The author is thankful to L. Mansfield for his considerable help in drawing and inserting the figures in this paper.

---

[1] The reduction from C^3 to C^2 was pointed out by Dombrowski.

§1 Conformal Structures

(1.1) Let V be a finite-dimensional real vector space. Two inner products g_1, g_2 are said to be conformal to each other if $g_1 = \lambda g_2$ for some positive real number λ. A conformal equivalence class of inner products on V is called a *conformal structure* on V. If $dim_{\mathbf{R}} V = n$ then the full group of linear automorphisms of V preserving a conformal structure on V is clearly $\approx R_+ \times O(n)$.

(1.2) Let M^n be a smooth [1] manifold. A *smooth conformal structure on* M^n is a smooth family of conformal structures on the tangent spaces $T_p(M^n)$, as p varies on M^n. Thus a conformal structure on M^n amounts to a covering $\{U_\alpha\}_{\alpha \in \Lambda}$ by open subsets of M^n and smooth Riemannian metrics g_α on U_α so that on the non-empty intersections $U_\alpha \cap U_\beta$ we have smooth positive functions

$$f_{\alpha\beta} : U_\alpha \cap U_\beta \longrightarrow R_+$$

with $g_\alpha = f_{\alpha\beta} g_\beta$. A smooth Riemannian metric on M^n of course defines its ambient smooth conformal structure, and evidently two Riemannian metrics have the same ambient conformal structure iff they are conformal to each other.

(1.3) Proposition A smooth conformal structure on M^n [2] is ambient to a smooth Riemannian metric.

Proof Let $\{U_\alpha,\ g_\alpha\}$ define the conformal structure as described above. Choose a smooth partition of unity $\{\varphi_\alpha\}$ subordinate to U_α. Then

$$g = \sum_\alpha \varphi_\alpha g_\alpha$$

is a Riemannian metric, and the conformal structure is ambient to it. *q.e.d.*

(1.4) Remark A partial improvement of **(1.3)** is that a real-analytic conformal structure is ambient to a real-analytic Riemannian metric. The proof is based on quite different ideas. (*Sketch of proof* : Let U_α, g_α, $f_{\alpha\beta}$ be as in **(1.2)** where now g_α and $f_{\alpha\beta}$ are real-analytic. Then $\log f_{\alpha\beta}$ defines a Čech 1-cocycle with the values

---

[1] Throughout "smooth" means C^∞. Sometimes we shall even omit this adjective. The instances where consideration of the degree of smoothness are important will be pointed out carefully later on.

[2] We always assume our manifolds to be Hausdorff and with a countable base for topology.

in the sheaf of germs of real-analytic functions. By complexifying M, using Stein neighborhoods, and using Cartan's theorem **A** and **B** one sees that this 1-cocycle is a 1-coboundary i.e. there exist real-analytic functions h_α defined on U_α such that $\log f_{\alpha\beta} = h_\alpha - h_\beta$ on non-empty intersections $U_\alpha \cap U_\beta$. Then

$$g \mid_{U_\alpha} = e^{h_\alpha} \cdot g_\alpha$$

is a globally defined real-analytic metric.)

(1.5) Let M^n be a manifold with a smooth conformal structure. A *conformal frame* at a point $p \in M$ is an orthogonal basis $\{e_1, \ldots, e_n\}$ of $T_p(M)$ such that e_i have equal length w.r.t. one (and hence any) compatible Riemannian metric. A conformal structure is *integrable* if each point $p \in M$ has a co-ordinate neighborhood with co-ordinates $\{x_1, \ldots, x_n\}$ such that $\left\{\frac{\partial}{\partial x_i}\right\}$ is a locally defined conformal frame. If the structure is integrable we shall call such co-ordinates *admissible*. It follows that if $\{x_1, \ldots, x_n\}$, $\{y_1, \ldots, y_n\}$ are two admissible co-ordinate systems defined on some common open subset U then $\left[\frac{\partial y_i}{\partial x_j}\right]$ defines a function : $U \longrightarrow R_+ \times O(n)$. In terms of an admissible co-ordinate system $\{x_1, \ldots, x_n\}$ a compatible Riemannian metric has the form

$$\lambda(x) \sum_i dx_i^2$$

where λ is a smooth $\mathbf{R}_+$ - valued function. So a manifold with integrable conformal structure may also be thought of as a *locally conformally Euclidean* manifold. For brevity we shall call such a manifold *conformally flat*.

(1.6) Proposition $\mathbf{S^n} = \{x \in \mathbf{E^{n+1}} \mid \| x \| = 1\}$ with its standard induced metric is conformally flat.

Proof Let $p \in \mathbf{S^n} \subseteq \mathbf{E^{n+1}}$ and π an n-plane $\subseteq \mathbf{E^{n+1}}$ parallel to $T_p(\mathbf{S^n})$ but different from $T_p(\mathbf{S^n})$. Given $q \in \mathbf{S^n} - \{p\}$ the line pq cuts π in exactly one point $\bar{q}$. The assertion is that the stereographic projection

$$\sigma \colon \mathbf{S^n} - \{p\} \longrightarrow \pi \ , \ \sigma(q) = \bar{q}$$

is a conformal homeomorphism.

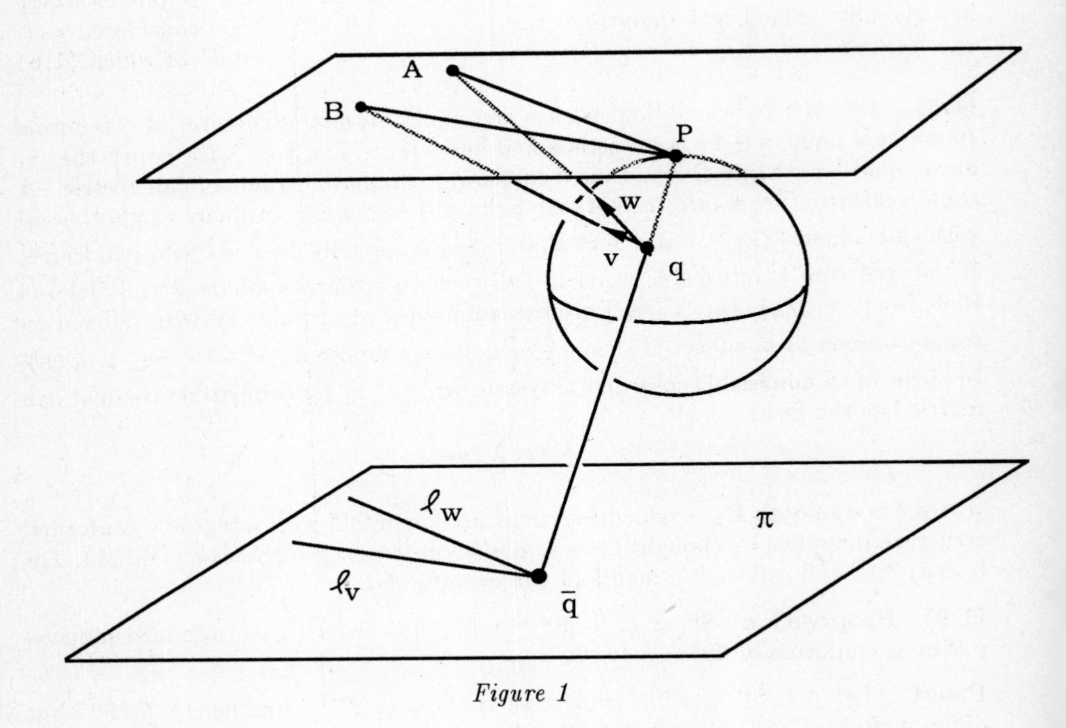

Figure 1

Take two oriented tangent vectors v, w at q. Then $\{p, v\}$, $\{p, w\}$ determine 2-planes cutting π in oriented lines l_v, l_w respectively. It suffices to show that the angles $\angle(v, w)$ and $\angle(l_v, l_w)$ are equal, and by continuity it suffices to show this for generic v and w. Now for generic v and w the tangent lines $\mathbf{R}v$, $\mathbf{R}w$ cut $T_p(\mathbf{S^n})$ at points A and B say. The oriented lines pA, pB are parallel to l_v, l_w respectively, so $\angle(pA, pB) = \angle(l_v, l_w)$. Also $| Ap |=| Aq |$ since these are lengths of the tangent segments from A to $\mathbf{S^n}$. Similarly $| Bp |=| Bq |$. So the triangles ApB, AqB are congruent. Hence

$$\angle(v,w) = \angle(pA,pB) = \angle(l_v,l_w).$$

Finally it is clear that σ can be expressed by rational functions in appropriate co-ordinates, so it is actually real-analytic. *q.e.d.*

(1.7) Remark We indicate another proof of **(1.6)** which in fact shows that the stereographic projection carries round p-spheres [1] into round p-spheres where in $\mathbf{E^n}$ considered as $\mathbf{S^n} - \{\infty\}$, a flat p-plane is allowed to be considered as a round p-sphere passing through ∞. This stronger result is "local" of which **(1.6)** is an "infinitesimal" case, cf. **(2.4)** below. Recall that an *inversion* in a round n-sphere $\mathbf{S^n}$ in $\mathbf{S^{n+1}} = \mathbf{E^{n+1}} \cup \{\infty\}$ is the map $P \mapsto Q$, *cf.* the figure, where $| OP | \cdot | OQ | = r^2$ with the understanding that $O \mapsto \infty$ and $\infty \mapsto O$.[2] It is well-known that an inversion carries a round p-sphere in $\mathbf{S^{n+1}}$ into a round p-sphere. Now consider the stereographic projection from the "north pole" N onto the tangent plane at the "south pole" S. The figure 2 shows that it coincides with the restriction to $\mathbf{S^n}$ of the inversion w.r.t. the sphere $\sum^n$ with center N and radius $| NS |$.

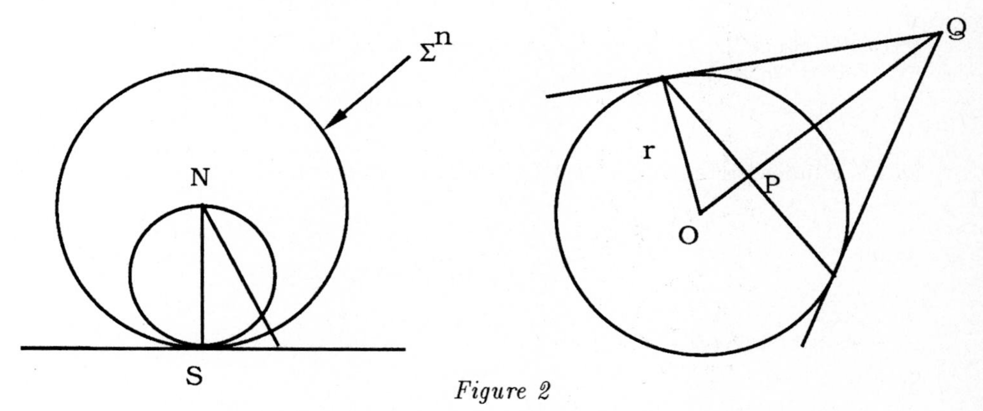

Figure 2

[1] A round p-sphere in $\mathbf{S^n} \subseteq \mathbf{E^{n+1}}$ is the transverse intersection of $\mathbf{S^n}$ with a flat $(p+1)$-plane.

[2] It is an intriguing fact that although this definition uses Euclidean notions, the inversion in $\mathbf{S^n} \subseteq \mathbf{S^{n+1}}$ is a Möbius notion. With more work it may be characterized in the Möbius category as follows: it is the unique non-identity Möbius transformation of $\mathbf{S^{n+1}}$ which commutes with all Möbius transformations leaving the given $\mathbf{S^n}$ invariant.

§2 Conformal Change of a Metric, Möbius structures

(2.1) Lemma. Let (M, g) be a Riemannian manifold, $\varphi: M \longrightarrow R$ a smooth function, and $\bar{g} = e^{2\varphi} g$ a conformal change of the metric. Let D, $\bar{D}$ denote the connections w.r.t. g and $\bar{g}$ respectively. Let $G = grad\,\varphi$ w.r.t. g, and define S on all vectorfields X,Y by

(2.1.1) $$\bar{D}_X Y = D_X Y + S_X Y$$

Then

(2.1.2) $$S_X Y = X\varphi Y + Y\varphi X - < X, Y > G.$$

Proof Since D, $\bar{D}$ are torsion-free we see that

$$S_X Y = S_Y X$$

and S is indeed a tensor i.e. it is C^∞-function-linear in both of its arguments. Next

$$\begin{aligned} 0 = \bar{D}_X \bar{g} = \bar{D}_X (e^{2\varphi} g) &= X(e^{2\varphi}) g + e^{2\varphi} \bar{D}_X g \\ &= e^{2\varphi} \{2X\varphi \cdot g + D_X g + S_X g\} \\ &= e^{2\varphi} \{2X\varphi \cdot g + S_X g\} \end{aligned}$$

So

$$(S_X g)(Y, Z) \overset{\text{def}}{=} - < S_X Y, Z > - < Y, S_X Z > = -2X\varphi < Y, Z >$$

Cyclically permuting X, Y, Z in the above equation and subtracting one equation from the sum of the other two and noting $S_X Y = S_Y X$ we obtain

$$\begin{aligned} < S_X Y, Z > &= X\varphi < Y, Z > + Y\varphi < Z, X > - Z\varphi < X, Y > \\ &= < X\varphi Y + Y\varphi X - < X, Y > G,\ Z > \end{aligned}$$

This implies (2.1.2). *q.e.d.*

(2.2) Let (M^n, g) be a Riemannian manifold and W^m a submanifold with the induced metric. Let D, ∇ be the connections defined by g and $g\,|_W$ respectively. As is well-known, if X, Y are smooth vectorfields tangential to W then

(2.2.1) $$D_X Y = \nabla_X Y + \alpha(X,Y)$$

where $\nabla_X Y$ coincides with the component of $D_X Y$ tangential to W and $\alpha(X,Y)$, *by definition* the normal component of $D_X Y$, is a symmetric bilinear form on the tangent bundle of W with values in its normal bundle, and is called the *second fundamental form* of W.

Let now $\bar{g} = e^{2\varphi} g$ be a conformal change of the metric as in **(2.1)** and define S and σ, *cf.* (2.1.1), by

(2.2.2) $$\begin{cases} \bar{D}_X Y = D_X Y + S(X,Y) \\ \bar{\nabla}_X Y = \nabla_X Y + \sigma(X,Y) \end{cases}$$

where $\bar{D}$ and $\bar{\nabla}$ are the connections of $\bar{g}$ and $\bar{g}\,|_W$ respectively. Along W decompose $G = grad\,\varphi$ (computed w.r.t. g) as

(2.2.3) $$G = G_0 + G_1$$

where G_0 resp. G_1 is the tangential resp. normal component. It follows by **(2.1)** that

(2.2.4) $$S(X,Y) = \sigma(X,Y) - \langle X, Y \rangle G_1$$

for all vectorfields tangent to W. Let $\bar{\alpha}$ denote the second fundamental form of W w.r.t. $\bar{g}$. Combining (2.2.1), (2.2.2) and (2.2.4) we get

(2.2.5) $$\begin{cases} \bar{\alpha}(X,Y) = \bar{D}_X Y - \bar{\nabla}_X Y \\ \qquad\qquad = D_X Y - \nabla_X Y + S(X,Y) - \sigma(X,Y) \\ \qquad\qquad = \alpha(X,Y) - \langle X, Y \rangle G_1 \end{cases}$$

If v is any normal vectorfield then if follows from (2.2.5) that the real-valued bilinear forms $\langle \bar{\alpha}, v \rangle$, $\langle \alpha, v \rangle$ differ by a scalar so the multiplicities of their eigenvalues are the same, and hence they are *conformal invariants* of W as a submanifold of M. In particular total umbilicity of a hypersurface (i.e. $m = n - 1$ and all eigenvalues of $\langle \bar{\alpha}, v \rangle$ are equal) is a conformally invariant notion.

(2.3) It is well-known that *for* $n \geq 3$ the totally umbilic hypersurfaces in the Euclidean n-space $\mathbf{E^n}$ are precisely the open subsets of the hyperplanes and the round $(n-1)$-spheres. A somewhat better expression of this fact is the statement : the totally umbilic hypersurfaces in the unit sphere $\mathbf{S^n}$ are precisely the open subsets of the round $(n-1)$-spheres, *cf.* (1.7). (This may be proved by the Riemann-geometric formulas, or else it also follows from the statement for $\mathbf{E^n}$, the stereographic projection, and the comments in **(2.2)**).

(2.3) This fact is at the basis of the rigidity in a conformally flat structure in dimension ≥ 3. For it follows by using the locally defined conformal maps into $\mathbf{E^n}$ (or $\mathbf{S^n}$) and **(2.2)** that on a conformally flat manifold M^n, $n \geq 3$, there is a distinguished family of submanifolds, namely the notion of "an open subset of a round $(n-1)$-sphere" (and hence by taking intersections also the notion of "an open subset of a round p-sphere", $1 \leq p \leq n-1$) has a meaning on M^n. In particular a conformal map among conformally flat n-manifolds, $n \geq 3$ preserves these pieces of round p-spheres.

(2.4) Proposition Let M^n be a conformally flat manifold, $n \geq 3$. The group of conformal diffeomorphisms of M^n precisely consists of those carrying round $(n-1)$-spheres into themselves.

Proof Let C_0 denote the group of conformal diffeomorphisms and C the group of those diffeomorphisms preserving the round p-spheres. As observed above, C_0 is contained in C. Let now $f \in C$. Our problem is to show that at a point p in M the differential f_{*p} is a homothety. Since the question is local we may assume that f fixes p and moreover a neighborhood of p is identified conformally with a neighborhood of the origin in E^n, so that p is identified with the origin. By using the polar decomposition of f_{*p} and composing by an element of $O(n)$ we may take f_{*p} to be a diagonal matrix

$$\Lambda = \begin{bmatrix} \lambda_1 & & 0 \\ & \ddots & \\ 0 & & \lambda_n \end{bmatrix}, \qquad \lambda_i > 0, \quad i = 1, 2 \ldots n.$$

in terms of the standard co-ordinate system. Now since f preserves $(n-1)$ - spheres, f_{*p} acting on T_p clearly has the same property. So

$$\lambda_1 = \lambda_2 = \ldots = \lambda_n$$

i.e. f_{*p} is a homothety.

q.e.d

(2.5) Let $M(n)$ denote the full group of diffeomorphisms of $\mathbf{S^n}$, $n \geq 2$, which carry round $(n-1)$-spheres into themselves. By **(2.4)** for $n \geq 3$ this group coincides with $C(\mathbf{S^n})$ the group of conformal diffeomorphisms. A basic fact in the theory of Riemann surfaces is that the same is true also for $n = 2$ — but structurally for quite different reasons from those in **(2.4)**! We shall see this in more detail later. We call $M(n)$ the *Möbius group in dimension n*, and its elements the *Möbius transformations.*

(2.6) Let M^n be a conformally flat manifold. Each admissible co-ordinate system on M^n gives a locally defined conformal map into $\mathbf{E^n}$ which we regard as $\mathbf{S^n}$ - {a point} via the stereographic projection. A *Möbius structure* on M^n is a maximal atlas of admissible co-ordinate charts such that the transition functions are restrictions of Möbius transformations. For brevity, we call a manifold with a Möbius structure a *Möbius manifold* and a local diffeomorphism preserving Möbius structures a *Möbius map.* It will be seen in §3 that for $n \geq 3$ a Möbius structure is the same as a conformally flat structure. For $n = 2$, an orientable conformally flat manifold is the same as a Riemann surface. On the other hand a *Möbius surface* is a finer notion than that of a Riemann surface. The study of Möbius structures on a given Riemann surface contains as a proper subset that of its "uniformizations" *à* la Fuchsian and Kleinian groups. The study of Möbius structures on the unit disk in $\mathbf{E^2}$ contains as a proper subset the theory of univalent functions. Thus the notion of a Möbius structure provides the geometric underpinnings of these grand classical theories, and provides a direction in which these theories admit a fruitful higher-dimensional generalization. In the context of Riemann surfaces, a Möbius structure is more commonly known as a $\mathbf{CP^1}$-structure, cf. [Gu].

(2.7) A class of Möbius manifolds arises as follows. Let Ω be an open subset of $\mathbf{S^n}$ and Γ a subgroup of M(n) which leaves Ω invariant and acts freely and properly discontinuously there. Then $\Gamma\backslash\Omega$ has a canonical Möbius structure. We shall call such manifolds *Kleinian.* The classical theory of Kleinian groups deals with subgroups Γ of M(2) which act properly discontinuously on some open non-empty subset of $\mathbf{S^2}$. Starting with such Γ one constructs in fact the largest open subset Ω of $\mathbf{S^n}$ on which Γ acts properly discontinuously.

§3 Liouville's Theorem

(3.1) Theorem (Liouville) Let U, V be open connected subsets of $\mathbf{S^n}$, $n \geq 3$ and $f: U \longrightarrow V$ be a conformal map. Then f is a restriction of a Möbius transformation $\tilde{f}$. Moreover $\tilde{f}$ is uniquely determined by f.

Proof The group M(n) contains $O(n+1)$ as well as the homotheties of $\mathbf{E^n}$ lifted to $\mathbf{S^n}$ by a stereographic projection. It easily follows that M(n) acts transitively on round n-balls and a closed round n-ball is conformal to the closed upper hemi-sphere.

Let B be a round n-ball contained in U. Then $f(B)$ is a round ball in V by **(2.4)**. Composing by an element of M(n) we may assume that f maps B into itself and B is the upper hemi-sphere. Let $r: \mathbf{S^n} \longrightarrow \mathbf{S^n}$ be the reflection in the equator ∂B. Then for $x \notin B$, the formula

(3.2.1) $$\tilde{f}(x) = rf(rx)$$

defines a conformal extension of f to $\mathbf{S^n}$. Moreover an extension of f to a small neighborhood of B is unique since any point in such a neighborhood is a point of intersection of circular arcs with a tail lying in $int\, B$ and f carries circular arcs to themselves. An easy connectedness argument now shows that f has a unique conformal extension to $\mathbf{S^n}$. So $\tilde{f}$ is this extension and $\tilde{f}$ coincides with f on the domain of definition of f. *q.e.d.*

(3.2) Remarks 1) There is a good deal of interest in determining the optimal smoothness requirements on the map for the validity of *Liouville's theorem.* For the moment the reader may take the above argument for C^∞-maps. Notice that since $\tilde{f}$ coincides with f in a neighborhood of B, $\tilde{f}$ would be C^∞ along with f. A careful examination of the proof will show that the above proof is valid for C^3-maps.[1] Usually in the proofs available in the literature one assumes f to be C^4. The

[1] The usual use of the Codazzi equation for establishing that the totally umbilic hypersurfaces in $\mathbf{E^n}$, $n \geq 3$ are pieces of flat planes or round spheres needs the C^3-hypothesis. However considering such hypersurface locally as a graph of a function it is easy to see that the proof is actually valid with the C^2-hypothesis.

notion of a conformal map makes sense for C^1-maps as well. But no simple proof of Liouville's theorm assuming f to be only C^1 is known. In the context of the theory of quasi-conformal homeomorphisms it is natural to define "conformal" to mean "1-quasiconformal". The validity of Liouville's theorem for 1-quasiconformal maps is due to Gehring, cf. [G].

2) The above proof for the case of C^2-maps also works in the pseudo-Riemannian analogues of *Liouville's theorem.*

3) Notice again the difference between the cases $n = 2$ and $n \geq 3$. In the above proof the hypothesis $n \geq 3$ enters through the use of **(2.4)**. Otherwise even for $n = 2$ the proof shows that a map defined on an open connected subset of $\mathbf{S^2}$ preserving circular arcs is a restriction of a Möbius transformation. On the other hand there is a rich family of conformal maps (which coincides with holomorphic or antiholomorphic maps with non-vanishing Jacobian) defined on open subsets of $\mathbf{S^2}$ (considered as the Riemann sphere). What is equally remarkable is that still the *globally defined* conformal maps of $\mathbf{S^2}$ preserve round circles.

4) The content of *Liouville's theorem* is that a conformally flat manifold of dimension $n \geq 3$ admits a canonical Möbius structure, cf.**(1.8)**. So the notions of a "conformally flat manifold" and a "Möbius manifold", are equivalent for $n \geq 3$.

§4. The Groups M(n) and M($\mathbf{E^n}$)

(4.1) The Euclidean space $\mathbf{E^n}$ has a cannonical Möbius structure. Let $M(\mathbf{E^n})$ denote the full group of Möbius diffeomorphisms of $\mathbf{E^n}$. Fix the origin $\vec{o}$. Clearly $\mathrm{M}(\mathbf{E^n})$ contains

(4.1.1) $\qquad \mathrm{O(n)} =$ the isometries fixing $\vec{o}$

(4.1.2) $\qquad R_+ =$ the homeotheties $\vec{x} \mapsto \lambda\vec{x}, \lambda \in \mathbf{R}_+, \vec{x} \in \mathbf{R^n}$

(4.1.3) $\qquad R^n =$ the translations.

The group of translations is normalized by O(n) and R_+ and the actions of O(n) and R_+ by conjugation are equivalent to their standard linear actions. Moreover these actions commute. So the group genereated by O(n), R_+, and R^n is isomorphic to the semidirect product

(4.1.4) $(\mathrm{O(n)} \cdot R_+) \triangleright R^n.$

Notice that this contains the full group of Euclidean isometries

(4.1.5) $\mathrm{E(n)} \stackrel{\mathrm{def}}{=} O(n) \triangleright R^n.$

(4.2) Theorem For $n \geq 2$ $M(\mathbf{E^n}) = (\mathrm{O(n)} \cdot R_+) \triangleright \mathbf{R^n}$.

Proof Let M_0 denote the right hand side, and write M for $\mathrm{M}(\mathbf{E^n})$. As remarked above $M_0 \subseteq$ M. Considering $\mathbf{S^n} = \mathbf{E^n} \cup \{\infty\}$, via the stereographic projection the group M may be identified with the subgroup of M(n) fixing ∞. Also a round (n-1) –sphere through ∞ corresponds to an (n-1) –plane in $\mathbf{E^n}$. It now follows from (2.4) that the elements of M carry (n-1)–planes into themselves, so by the socalled fundamental theorm of projective geometry valid for $n \geq 2$ it follows that M is contained in the group $Aff(n)$ of affine transformation of $\mathbf{E^n}$.

Since

(4.2.1) $Aff(n) = GL(n) \triangleright R^n$

(4.2.2) $\mathrm{M}(\mathbf{E^n}) \cap GL(n) = O(n) \cdot R_+$

We see that $M = M_0$. *q.e.d.*

(4.3) Remark For $n \geq 3$, by Liouville's theorem $\mathrm{M}(\mathbf{E^n})$ coincides with the full group of conformal diffeomorphisms. This fact is also valid for $n = 2$ and is essentially equivalent to the fact that M(2) coincides with $C(\mathbf{S^2})$.

(4.4) The theorem shows that $\mathrm{M}(\mathbf{E^n})$ is a Lie group and consists of real-analytic transformations. Since $\mathrm{M}(\mathbf{E^n})$ is the isotropy subgroup of M(n) at the point $\{\infty\}$ it follows that M(n) is also a Lie group and consists of real-analytic transformations. This implies that the transition functions of a Möbius structure and the maps preserving a Möbius structure are also real-analytic. Note also that

(4.4.1) $\dim \mathrm{M}(\mathbf{E^n}) = \frac{1}{2}(n)(n-1) + 1 + n = \frac{1}{2}(n^2 + n + 2).$

(4.4.2) $\dim \mathrm{M(n)} = \dim \mathrm{M}(\mathbf{E^n}) + n = \frac{1}{2}(n+1)(n+2).$

Each of $\mathrm{M}(\mathbf{E^n})$ and M(n) have two components. The identity component is orientation-preserving and the other orientation-reversing.

(4.5) The linear model for M(n). Consider $\mathbf{R}^{n+2}$ with the quadratic form of signature (1, n+1).

(4.5.1) $Q(x) = x_0^2 - x_1^2 - \cdots - x_{n+1}^2.$

Let

(4.5.2) $\quad O(Q) = O\ (n,\ 1) =$ the group of linear isometries of $(\mathbf{R}^{n+2},\ Q)$.

Then dim $O(Q) = \frac{(n+1)(n+2)}{2}$ and has 4 components. Also $O(Q)$ preserves the quadric

(4.5.3) $$X(Q) = \{x \in \mathbf{R}^{n+2} \mid Q(x) = 0\},$$

which is a right-spherical cone in $\mathbf{R}^{n+2}$ with vertex at the origin. The subgroups of $O(Q)$ which preserves the half-space $x_0 > 0$ contains 2 out of 4 components. For the want of a better notation we denote this subgroup by

(4.5.4) $O_{\frac{1}{2}}(Q) = \{g \in O(Q) | g \text{ preserves } x_o > 0\}$

Now $O_{\frac{1}{2}}\ (Q)$ also acts on the space of rays in $X(Q)$ through the origin lying in the half space $x_0 > 1$. This space of half-rays in turn may be identified with

(4.5.5) $$\{x \in \mathbf{R}^{n+2} \mid Q(x) = 0, x_0 = 1\} = \mathbf{S^n}.$$

The round (n-1)-spheres in $\mathbf{S^n}$ are precisely the transverse intersections of the (n+2) - planes in $\mathbf{R}^{n+2}$ through the origin. It follows that $O_{\frac{1}{2}}(Q)$ acting on $\mathbf{S^n}$ as above carries round (n-1) -spheres into themselves. So by (2.4)

(4.5.6) $$M(n) \approx O_{\frac{1}{2}}\ (Q).$$

This identifies M(n) as a Lie group.

§5. The Connection with Hyperbolic Geometry

(5.1) Let

(5.1.1) $$\mathbf{D^n} = \{x \in \mathbf{E^n} \mid |x| < 1\}.$$

(5.1.2) $$\mathbf{H^n} = \{x = (x_1 \ldots x_n) \in \mathbf{E^n} \mid x_n > 0\}.$$

We equip $\mathbf{D^n}$, $\mathbf{H^n}$ with the Möbius structures induced from $\mathbf{E^n}$. Let M($\mathbf{D^n}$), M($\mathbf{H^n}$) denote the full groups of Möbius–structure–preserving diffeomorphisms of $\mathbf{D^n}$ and $\mathbf{H^n}$ respectively.

(5.2) Proposition $\mathbf{D^n}$ and $\mathbf{H^n}$ are equivalent as Möbius manifolds. M($\mathbf{H^n}$) is transitive on $\mathbf{H^n}$.

Proof Via a streographic projection we may regard $\mathbf{D^n}$ as the "southern" hemisphere in $\mathbf{S^n}$. By an appropriate rotation it is equivalent to the "eastern" hemisphere. Now a streographic projection from the "north pole" would map the eastern hemisphere onto a half-space in $\mathbf{E^n}$.

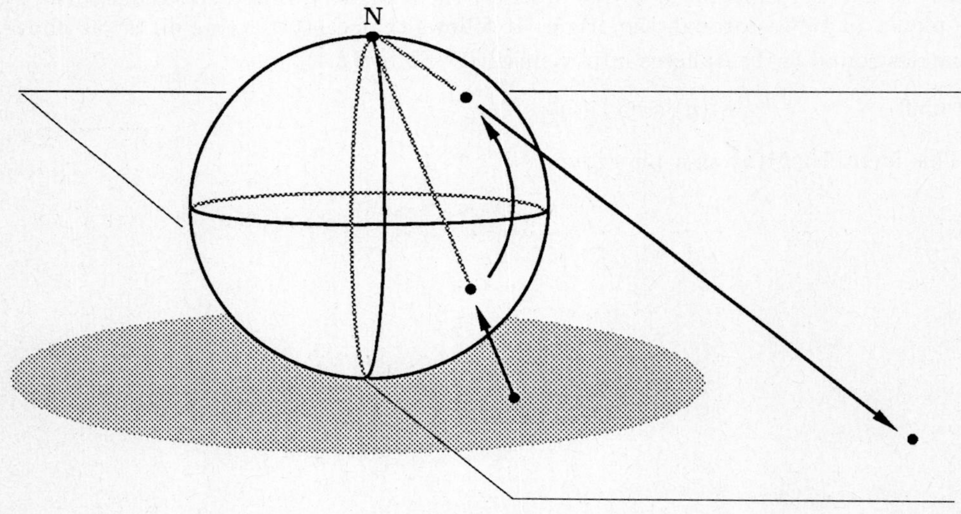

Figure 3

Since all half-spaces in $\mathbf{E^n}$ are mutually isometric we see that $\mathbf{D^n}$ and $\mathbf{H^n}$ are equivalent as Möbius manifolds. (This correspondence between $\mathbf{D^n}$ and $\mathbf{H^n}$ is classically known as a *a Cayley transform.*)

Next the maps

(5.2.1) $$x \mapsto x + a,\ a = (a_1 \dots a_{n-1}, 0)$$

(5.2.2) $$x \mapsto \lambda x\,, \quad \lambda > 0$$

clearly belongs to $M(\mathbf{H^n})$ and it is easy to see that the group generated by these maps is transitive on $\mathbf{H^n}$. *q.e.d.*

(5.3) A remarkable property of $\mathbf{D^n}$, not shared by $\mathbf{S^n}$ or $\mathbf{E^n}$ is that

Proposition $\mathbf{D^n}$ admits a $M(\mathbf{D^n})$ - invariant Riemannian metric. This metric is complete and has constant negative curvature.

Proof It is convenient to consider $\mathbf{D^n}$ as the hemi–sphere in $\mathbf{S^n} \subseteq \mathbf{E^{n+1}}$:

(5.3.1) $$\mathbf{D^n} = \{x \in \mathbf{E^{n+1}} \,|\; |x| = 1,\ x_{n+1} > 0\}.$$

The stereographic projection from $P = (0, 0 \dots 0, 1)$

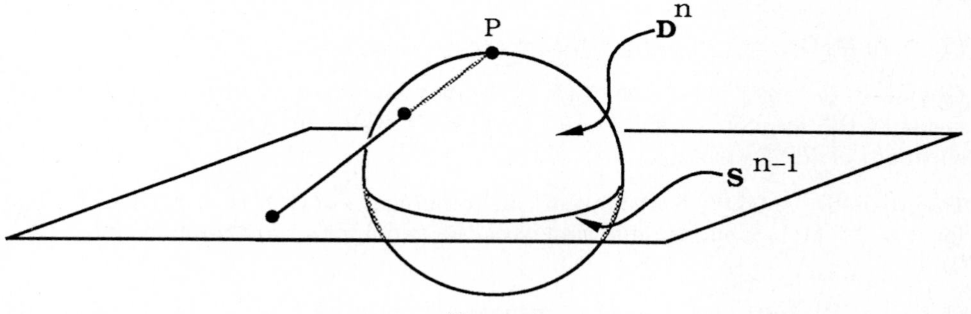

Figure 4

onto $\mathbf{E^n} = \{x \in \mathbf{E^{n+1}} \mid x_{n+1} = 0.\}$ maps $\partial\mathbf{D^n}$ onto the unit sphere $\mathbf{S^{n-1}}$ in $\mathbf{E^n}$. Now let H denote the isotropy subgroup of M(n) at P. Then H may be identified with $M(\mathbf{E^n})$. So the isotropy subgroup K of $M(\mathbf{D^n})$ at P equals $H \cap M(\mathbf{D^n})$. So K may be identified with the subgroup of $M(\mathbf{E^n})$ which leaves $\mathbf{S^{n-1}}$ invariant. From the description of $M(\mathbf{E^n})$ in (4.2) it follows easily that $K \approx O(n)$. In particular K is *compact.* So taking a K–invariant metric on the tangent space $T_p(\mathbf{D^n})$ and translating by $M(\mathbf{D^n})$, we obtain an $M(\mathbf{D^n})$ - invariant Riemannian metric on $\mathbf{D^n}$. Since O(n) is transitive on 2-dimensional subspaces of $T_p(\mathbf{D^n})$ it follows that the metric has a constant Riemannian curvature, and as is true for any homogenous

Riemannian metric, this metric is complete. By a basic theorem in differential geometry, $\mathbf{D^n}$ *w.r.t.* this metric after scaling if necessary is isometric to either $\mathbf{S^n}$, or $\mathbf{E^n}$, or the hyperbolic space i.e. a complete simply connected Riemannin manifold with constant curvature -1. But $\mathbf{D^n}$ cannot be $\mathbf{S^n}$ or $\mathbf{E^n}$, for these do not admit metrics invariant under their full groups of Möbius diffeomorphisms since the image of their isotropy representations is clearly not contained in the orthogonal group. So $\mathbf{D^n}$ or equivalently $\mathbf{H^n}$ must be isometric to the hyperbolic space (up to a scaling factor). *q.e.d.*

(5.4) We shall now identify M($\mathbf{D^n}$) as a Lie Group, for $n \geq 2$. We may identify $\mathbf{D^n}$ with a round ball B in $\mathbf{S^n}$. So by Liouville's theorem for $n \geq 3$ and by the comments thereafter for $n = 2$ we have

(5. 4. 1) $\mathrm{M}(\mathbf{D^n}) = \{g \in M(n) | g(B) = B\}$.

Now it is convenient to use the linear model, cf. (4. 5). In that model we may take

(5. 4. 2) $\mathbf{D^n} = \{x \in R^{n+2} \,|\, Q(x) = 0, x_0 = 1, x_{n+1} > 0\}$

So,

(5. 4. 3) $\mathrm{M}(\mathbf{D^n}) = \{g \in O_{1/2}(Q) \,|\, g \textit{ preserves } x_{n+1} > 0\}$

Obviously the right hand side of (5. 4. 3) may be identified with $O_{1/2}(Q_0)$ where Q_0 is of the form $x_0^2 - x_1^2 - \ldots - x_n^2$ on R^{n+1} which in turn is considered as a subspace of R^{n+2} with $x_{n+1} = 0$. Thus,

Proposition M($\mathbf{D^n}$) $\approx$ two out of four components of O(1, n) $\approx M(n-1)$ where for n = 2 † M(1) is interpreted simply as the two appropriate components of O(1, 2).

(5.5) We now view the above situation from a different angle. Consider $\mathbf{S^n}$ as $\partial\mathbf{D^{n+1}}$. Then (5. 4) means that the Möbius group, M(n), uniquely extends to $\mathbf{D^{n+1}}$, and on $\mathbf{D^{n+1}}$ acts as a full group of isometries of $\mathbf{D^{n+1}}$ with respect to the hyperbolic metric. This sets up a fundamental relationship between conformal geometry and hyperbolic geometry. On the one hand, a hyperbolic manifold, i.e., a (not necessarily complete) Riemannian manifold of constant negative curvature has a canonical Möbius structure. On the other hand, a Möbius manifold is an "ideal boundary" of some hyperbolic manifold. Anticipating later developments we make this last statement more precise in two ways.

---

† It turns out that M(2) coincides with the full group of conformal diffeomorphisms of $\mathbf{S^2}$.

(5.6) **Proposition**: Let $M^n = \Gamma\backslash\Omega$ be a Kleinian manifold such that Γ is torsionfree. Then, there exists a manifold N^{n+1} with boundary $= M^n$ such that int N^{n+1} admits a complete hyperbolic metric.

Proof: By hypothesis we have $M^n = \Gamma\backslash\Omega$ where Ω is an open non-empty subset of $\mathbf{S^n}$, and $\Gamma \leq M(n)$, such that Γ leaves Ω invariant and acts freely and properly discontinuously there. Consider

(5. 6. 1) $\tilde{N} = \Omega \cup \mathbf{D^{n+1}}$.

As explained in (5. 4) the action of Γ on Ω extends cannonically on $\tilde{N}$, and in fact Γ acts as a group of isometries on $\mathbf{D^{n+1}}$. Since Γ acts properly discontinuously on Ω it is discrete (in the compact-open topology) as a group of homeomorphisms of $\mathbf{D^{n+1}}$. Now a discrete group of isometries acts properly discontinuously. Since Γ is torsionfree, its action on $\mathbf{D^{n+1}}$ is both free and properly discontinuous – so $\Gamma\backslash\mathbf{D^{n+1}}$ is a complete hyperbolic manifold. A subtle point which will not be justified here is that the action of Γ on $\tilde{N}$ is also properly discontinuous. So $N^{n+1} = \Gamma\backslash\tilde{N}$ is a manifold with boundary and obviously it has the required properties. *q.e.d.*

(5.7) We now give a second construction of a manifold N^{n+1} with "ideal boundary", a given Möbius manifold M^n so that int N is a hyperbolic manifold. Its idea goes back to Nielsen and it may be called "*a convex hull construction*". Let $p : \tilde{M}^n \longrightarrow M^n$ be the universal covering projection of a Möbius manifold M^n - so $M^n \approx \Delta\backslash\tilde{M}^n$ where $\Delta \approx \pi_1(M)$ is the deck-transformation group. We assume that Δ is torsionfree. Pull back the Möbius structure from M^n to $\tilde{M}^n$. Let us also assume that $\tilde{M}^n$ is non-compact. It will be clear after the discussion of the "development map" in §7 that $\tilde{M}^n$ has the property that two round n-balls in $\tilde{M}^n$ are either disjoint or else they intersect in a "spherical lens", cf. the figure below.

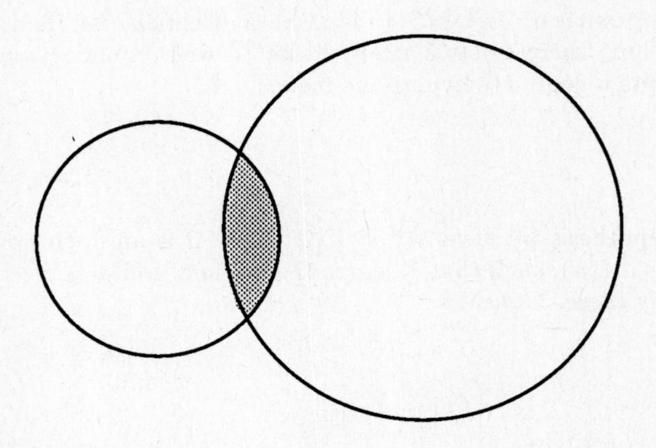

Figure 5

Now represent $\tilde{M}^n$ as the union $\cup_{i\epsilon I} B_i^n$ where B_i runs over all the round balls in $\tilde{M}^n$. With each B_i we can associate a half round $(n+1)$-ball H_i^{n+1} containing B_i so that the hemi-spherical boundary of H_i^{n+1} cuts B_i orthogonally, cf. (5.7.2)

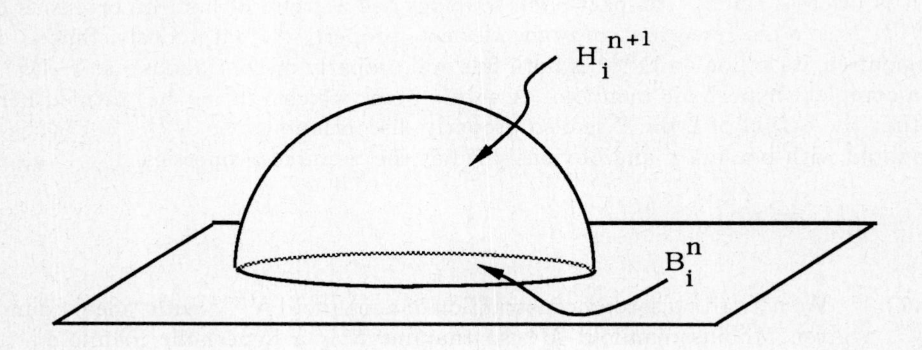

Figure 6

Let $\cup_{i\epsilon I} H_i^{n+1}$ be a disjoint union. We construct an $(n+1)$-dimensional manifold

(5.7.3) $$\tilde{N}^{n+1} = \cup_{i\epsilon I} H_i^{n+1} / \sim,$$

where $\sim$ is an equivalence relation defined as follows. If $B_i \cap B_j$ is empty then no point of H_i^{n+1} is identified with a point in H_j^{n+1}. Suppose $B_i \cap B_j$ is non-empty. Then there is a Möbius embedding

$$\delta : B_i \cup B_j \longrightarrow \mathbf{S^n}. \tag{5.7.4}$$

Considering $\mathbf{S^n}$, as $\partial\mathbf{D^{n+1}}$ let $\bar{H}_i^{n+1}$, $\bar{H}_j^{n+1}$ be the half round (n+1)-balls contained in $\bar{\mathbf{D}}^{n+1}$ such that $\bar{H}_i^{n+1} \cup \mathbf{S^n} = B_i$ and the hemispherical boundary of H_i cuts S^n orthogonally. Obviously δ extends to a surjective map

$$\delta_1 : H_i \cup H_j \longrightarrow \bar{H}_i \cup \bar{H}_j\,. \tag{5.7.5}$$

We identify a point in H_i with a point in H_j if they have the same δ-image. It will follow from the theory of development that this identification is independent of the choice of δ. Then $\sim$ is simply the equivalence elation generated by this process. It is not difficult to see that $\tilde{N}^{n+1}$ is a manifold with boundary $\approx M^n$. Moreover after identifying $\partial\tilde{N}^{n+1}$ with $\tilde{M}^n$, the action of Δ extends to $\tilde{N}^{n+1}$ and is free and poperly discontinuous. Then $N^{n+1} = \Delta\backslash\tilde{N}^{n+1}$ is a manifold with boundary M^n. This N^{n+1} is called the *Nielsen convex hull* of M^n.

§6 Constructions of Möbius Manifolds

(6.1) A 2-dimensional smooth Riemannian manifold M^2 is conformally flat. This is the content of the famous"existence of isothermal co-ordinates." A priori it is not clear that M^2 admits a compatible Möbius structure. However by the uniformization theorem M^2 admits a complete metric of constant curvature conformal to the given one. This equips M^2 with one choice of in fact a canonical Möbius structure. On the other hand, except for $M^2 = \mathbf{S^2}$, in all other cases M^2 admits an exceedingly rich family of other Möbius structures. For example, for $M^2 = \mathbf{E}^2$ or $\mathbf{D^2}$, considered as Riemann surfaces, any meromorphic function with nowhere vanishing jacobian equips M^2 with a Möbius structure.

In all dimensions we now look for constructions of Kleinian manifolds.

(6.2) $\Omega = \mathbf{S^n}$: Since Ω is compact a group Γ acting properly discontinuously must be finite. From (4.5) we see that O(n+1) is a maximal compact subgroup of M(n). By a standard result in Lie theory Γ is conjugate to a subgroup of O(n+1). So $\Gamma\backslash\Omega$'s obtained this way are precisely the conformal types of sperical space forms.

(6.3) $\Omega = \mathbf{S^n}$ - $\{a\ point\}$ which is Möbius equivalent to $\mathbf{E^n}$. From the description of $M(\mathbf{E^n})$ we see that an element in $M(\mathbf{E^n}) - E(n)$ is of infinite order and has a fixed point, so it cannot be contained in any Γ acting freely and properly discontinuously on $\mathbf{E^n}$. In other words if $\Gamma \leq M(\mathbf{E^n})$ which acts freely and properly discontinuously on $\mathbf{E^n}$ then in fact $\Gamma \leq E(n)$. So $\Gamma\backslash\mathbf{E^n}$ are precisely the conformal types of Euclidean space-forms.

(6.4) $\Omega = \mathbf{S^n} - \{2\ points\}$: Alternately we may consider Ω as $\mathbf{E^n} - \{0\}$. Clearly $M(\Omega)$ has a subgroup G of index 2 which fixes 0 and ∞, $G \approx R_+ \times 0(n)$, and G acts transitively on Ω with isotropy subgroup at a point $\approx O(n-1)$. Since this is compact we see that Ω admits an $M(\Omega)$ - invariant Riemannian metric. In fact in terms of polar co-ordinates we may take this metric to be

$$\frac{dr^2}{r^2} + d\sigma^2_{n-1}, \tag{6.4.1}$$

where $d\sigma^2_{n-1}$ is the standard metric on $\mathbf{S^{n-1}}$ making Ω isometric to $\mathbf{R} \times \mathbf{S^{n-1}}$. It now follows that a discrete subgroup of $M(\Omega)$ acts properly discontinuously on Ω. It is easy to see that a discrete subgroup of G is either finite or contains a subgroup $\Phi \approx \mathbf{Z}$ of finite index generated by an element of the form $\lambda \cdot A, \lambda > 1$ and $A \epsilon O(n)$ where A = e or is of infinite order. Then $\Phi\backslash\Omega$ is diffeomorphic to $\mathbf{S^{n-1}} \times \mathbf{S^1}$. Both λ and the rotation angles of A are invariants of the Möbius structure on $\Phi\backslash\Omega$. A compact quotient of Ω by $\Gamma \leq M(\Omega)$ acting freely and properly discontinuously is called a *Hopf manifold.* As noted above, a Hopf manifold is finitely covered by a manifold diffeomorphic to $\mathbf{S^{n-1}} \times \mathbf{S^1}$.

A 2-dimensional Hopf manifold is diffeomorphic to a torus or a Klein bottle. Its Möbius structure is different from the canonical one corresponding to the Euclidean metric compatible with its conformal structure.

(6.5) $\Omega = \mathbf{S^n} - \mathbf{S^p}$, $1 \leq p \leq n-1$: Here $\mathbf{S^p}$ denotes a round p-sphere in $\mathbf{S^n}$. Taking a stereographic projection from a point in $\mathbf{S^p}$ we see that Ω is Möbius equivalent to $\mathbf{E}^n - \mathbf{E}^p$, where $\mathbf{E}^p$ is a linear p-dimensional subspace.

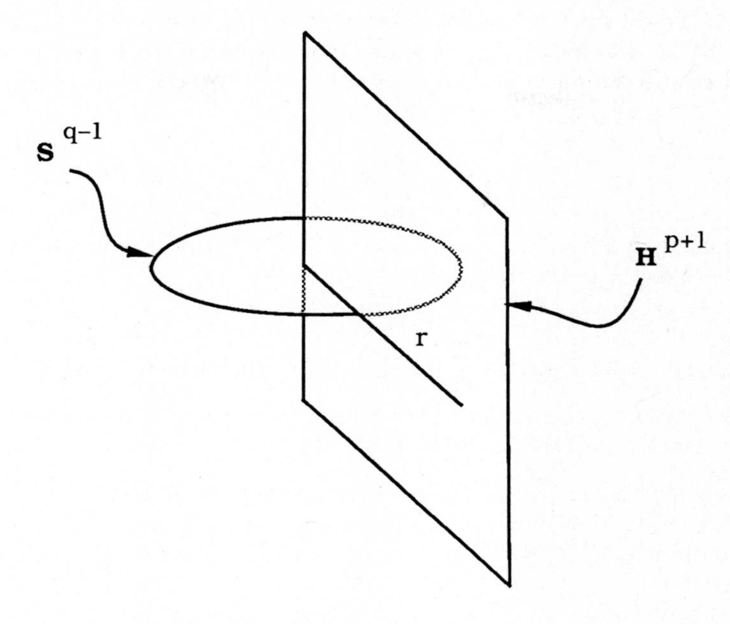

Figure 7

Then the subgroup

(6.5.1) $\{g \in M(\mathbf{E^n}) | \, g(\mathbf{E^p}) = \mathbf{E^p}\}$

contains translations parallel to $\mathbf{E^p}$, rotations based at a point in $\mathbf{E^p}$ and leaving $\mathbf{E^p}$ invariant, and the homotheties based at a point in $\mathbf{E^p}$. This subgroup is already transitive on Ω. To determine $M(\Omega)$ precisely, it is convenient to go to the linear model,cf. (4.5). Let V be a (p+2) - dimensional subspace of $\mathbf{R}^{n+2}$ such that $Q\,|_V$ has type (1, p+1). Then V intersects $\mathbf{S^n}$ in a round p-sphere. So

(6.5.2) $M(\Omega) = \{g \in M(n) | \, g(V) = V\}.$

If W is Q-orthogonal to V in $\mathbf{R}^{n+2}$ then $\mathbf{R}^{n+2} = V \oplus W$ and $Q\,|_W$ has type (0, n-p). Clearly

(6.5.3) $M(\Omega) \approx 0_{\frac{1}{2}}(p+1, 1) \times O(n-p).$

The isotropy subgroup of $M(\Omega)$ at a point is seen to be $\approx O(p+1) \times O(n-p-1)$.

Since this group is compact Ω admits an $M(\Omega)$ - invariant Riemannian metric. In fact if we take the standard cartesian co-ordinates $(x_1 \dots x_p, y_1 \dots y_q)$, $p+q=n$, in $\mathbf{E}^n$ so that $\mathbf{E}^p = \{(x_1 \dots x_p, 0 \dots 0)\}$ then on $\mathbf{E}^n - \mathbf{E}^p$ we have $y_1^2 + \cdots + y_q^2 \neq 0$. Choosing the polar co-ordinates *w.r.t.* $\{y_1 \dots y_q\}$ we can write

$$\sum_{i=1}^{q} dy_i{}^2 = dr^2 + r^2 d\sigma_{q-1}^2,$$

where $d\sigma_{q-1}^2$ is the metric on $\mathbf{S^{q-1}}$.

So

(6.5.4) $\quad \frac{1}{r^2}\{\sum dx_i{}^2 + \sum dy_i{}^2\} = \frac{1}{r^2}\{\sum dx_i{}^2 + dr^2\} + d\sigma_{q-1}^2.$

We leave it to the reader to verify that this metric is invariant under $M(\Omega)$ making Ω isometric to $\mathbf{H^{p+1}} \times \mathbf{S^{q-1}}, p+q=n$.

It follows in particular that a discrete subgroup of $M(\Omega)$ acts properly discontinuously on Ω. Let us now consider a discrete subgroup Γ which is also torsion-free. Identifying $M(\Omega)$ with $O_{\frac{1}{2}}(1, p+1) \times O(n-p)$ we write an element of $M(\Omega)$ as a pair

(6.5.5) $\quad g = (h,k), \;\; h \in O_{\frac{1}{2}}(1, p+1), \;\; k \in O(n-p).$

Since $\Gamma \cap \{e \times O(n-p)\}$ is compact and discrete and Γ is assumed to be torsion-free we see that

(6.5.6) $\quad \Gamma \cap \{e \times O(n-p)\} = \{e\}.$

So $\Gamma \ni g = (h,k) \longmapsto h$ is an isomorphism, and clearly the image Γ_1 of Γ in $O_{\frac{1}{2}}(1, p+1)$ is also discrete. In other words, $\Gamma_1 \approx \pi_1$(a hyperbolic manifold of dimension p+1). We can now write

(6.5.7) $\quad \Gamma = \{(h, \rho(h)) | h \in \Gamma_1, \, \rho : \Gamma_1 \to O(n-p)\}$

where ρ is some representation. The map

(6.5.8) $\quad \Gamma \backslash \Omega \to \Gamma_1 \backslash \mathbf{H}^{p+1}$

equips $\Gamma \backslash \Omega$ with a structure of a locally flat fiber bundle with fiber $\mathbf{S^{n-p-1}}$ and structure group Γ_1 acting on $\mathbf{S^{n-p-1}}$ via ρ. Conversely given $\Gamma_1 \leq O_{\frac{1}{2}}(p+1, 1)$ acting freely and properly discontinuously on $\mathbf{H^{p+1}}$, and a representation $\rho : \Gamma_1 \to O(n-p)$ we can construct Γ as in (6.4.7) and thus a Möbius manifold $\Gamma \backslash \Omega$.

This construction gives a large class of Möbius manifolds with varied possibilities for topological types. Moreover the possibilities of variations of ρ indicate the possibilities for deformations of a Möbius structure.

A comment on the hypothesis that Γ was assumed to be torsion-free in the above construction: From (6.4.3) we see that $M(\Omega)$ admits a faithful linear representation. So if Γ is finitely generated then by a well-known property of linear groups, Γ admits a torsion-free subgroup of finite index. So any $\Gamma\backslash\Omega$ with Γ finitely generated, (in particular if $\Gamma\backslash\Omega$ is compact), admits a finite covering which is a locally flat, orthogonal $\mathbf{S}^{\mathbf{n-p-1}}$–bundle over a hyperbolic manifold of dimension p+1.

Finally notice the following special cases. If p = n-1 then $\mathbf{S}^{\circ} = \{2\ points\}$ and $\mathbf{S}^{\mathbf{n}} - \mathbf{S}^{\mathbf{n-1}}$ has two components each Möbius equivalent to $\mathbf{H}^{\mathbf{n}}$. If p = n-2 then Ω is not simply connected. Its universal cover is Möbius equivalent to the Riemannian product $\mathbf{H}^{\mathbf{n-1}} \times \mathbf{R}$.

(6.6) The connection with Kleinian groups: Let Γ be a discrete subgroup of $M(n)$. When n = 2, such Γ is classically called a *Kleinian group*, and when n = 1, a *Fuchsian group*. We may regard Γ as acting on $\mathbf{H}^{\mathbf{n+1}}$ or $\mathbf{D}^{\mathbf{n+1}}$, cf. §5, as a group of hyperbolic isometries. For $p \in \mathbf{D}^{n+1}$ let Λ denote the set of accumulation points in $\mathbf{D}^{n+1} \cup \mathbf{S}^{\mathbf{n}}$, since Γ is a discrete group of isometries on $\mathbf{D}^{n+1}$ one readily sees that $\Lambda \subseteq \mathbf{S}^{\mathbf{n}}$. A basic observation which we do not prove here is that Γ acts properly discontinuously on $\Omega = \mathbf{S}^{\mathbf{n}} - \Lambda$, and in fact on $\mathbf{D}^{n+1} \cup \Omega$. If Ω is not empty, and if Γ acts freely on Ω then $\Gamma\backslash\Omega$ is a (not necessarily connected) Kleinian manifold.

The case n=2, and Ω connected occurs in the extensive studies by Köebe, recently explained and extended in a beautiful series of papers by Maskit $[M]_1$, based on his purely topological understanding of planar coverings, $[M]_3$.

In the case n = 2, Γ finitely generated, and Ω non-empty but not necessarily connected Ahlfors [A] proved a fundamental theorem: $\Gamma\backslash\Omega$ has only finitely many components, each component is biholomorphic to a compact Riemann surface with at most finitely many punctures. If $\Omega \neq \mathbf{S}^{\mathbf{2}}$, $\mathbf{S}^{\mathbf{2}} - \{a\ point\}$, or $\mathbf{S}^{\mathbf{2}} - \{2\ points\}$ then each component of $\Gamma\backslash\Omega$ is obviously of hyperbolic type. cf. also Bers [B], Greenberg [Gr]. Classically Γ is said to be *non – elementary* if $\Gamma\backslash\Omega$ is of hyperbolic type, i.e. when $\#\Lambda \geq 3$. An end of the 3-manifold $\Gamma\backslash\mathbf{D}^3$ with finite hyperbolic volume is called a *cusp*. (This notion can be given a meaning whether or not Γ is torsion-free.) Bers [B] improved Ahlfors' theorem by giving an estimate for the hyperbolic area of $\Gamma\backslash\Omega$, (Γ non-elementary, finitely generated) in terms of the number of generators of Γ. Whether Ω is non-empty or not Sullivan [S] proved that $\Gamma\backslash\mathbf{D}^3$ has only finitely many cusps. He also gave an estimate for their number in terms of the number of generators of Γ (Sullivan's proof and estimates are valid only for Γ, torsion free. For Γ not necessarily torsion-free these results are extended by Kra [K]$_3$). The underlying topological statements in these results of Ahlfors, Bers, and sullivan for Γ torsion-free were explained purely in terms of 3-dimensional topology by Kulkarni and Shalen [KS] with further improvements by McCullough [M] and McCullough - Feighn [FM] .

In a by-now well-known, but far from well-digested work Thurston has shown that "most" compact 3-manifolds can be represented as $\Gamma\backslash\mathbf{D}^3$; if the manifold is closed, or has only torus boundary components, this structure, if it exists, is unique by Mostow [Mo]. These various interconnections is a rich source of food for thought for the geometers.

(6.7) Schottky manifolds: This is a nice class of Kleinian manifolds which classically (for n=2) motivated the general Kleinian constructions explained in (6.5). Let $\{C_i, C'_i\}$ $i = 1, 2, \ldots g$ be a family of (n-1) - spheres in S^n which bound n-disks $\{D_i, D'_i\}$ say, so that all D_i's and D'_i's are mutually disjoint. Suppose there exist $\gamma_i \in M(n)$ such that $\gamma_i(D_i) = \mathbf{S^n} - D'_i$. It is easy to see that each γ_i has two fixed points, one in D_i and the other in D'_i. (If C_i, C'_i are *round* (n-1) - spheres such γ_i surely exist.) If g=1 then $\Lambda = \{$ the fixed points of $\gamma_1\}$, and $< \gamma_1 > \backslash\{\mathbf{S^n} - \Lambda\}$ is a Hopf manifold, cf. (6.3). Consider the case g = 2.

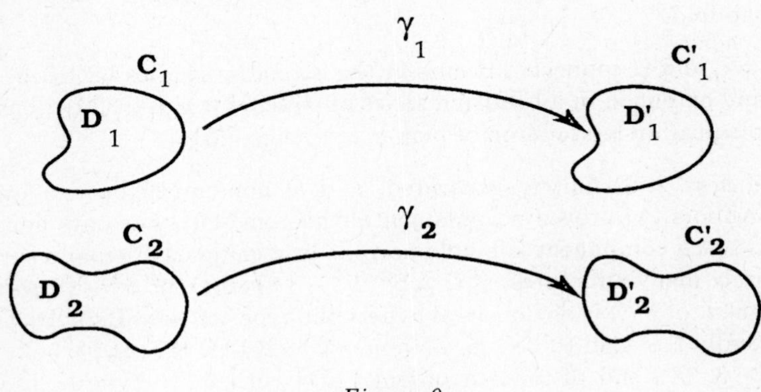

Figure 8

The images of $\mathbf{S^n} - D_1$ under $\gamma_1, \gamma_1^2, \ldots$ would look like

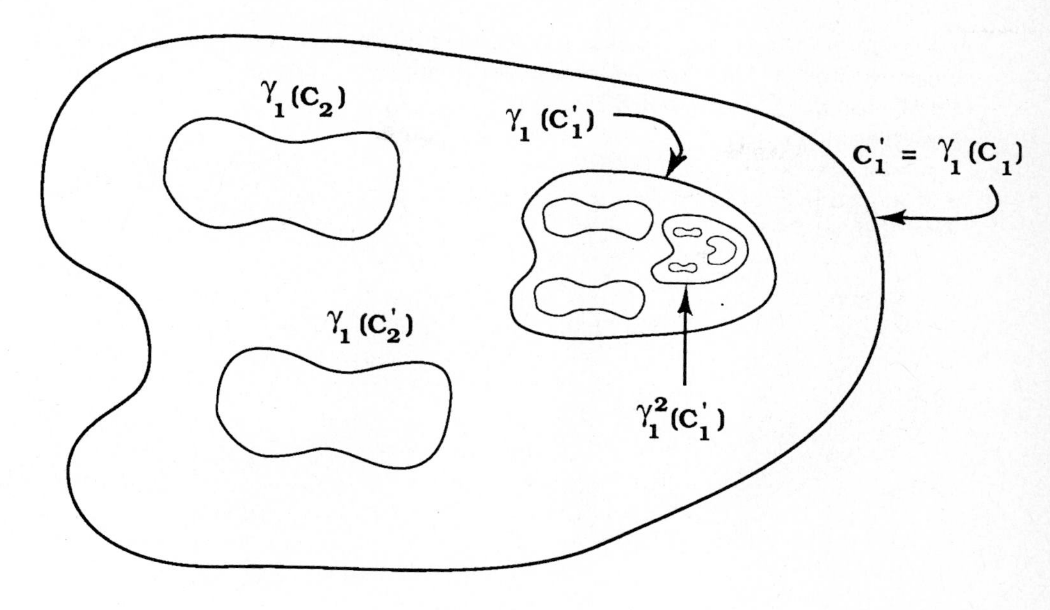

Figure 9

From these pictures for g =2 it is fairly clear that $\mathbf{S^n} - \bigcup_{i=1}^{g}\{D_i \cup D'_i\}$ is a fundamental domain for $\Gamma = \langle\gamma_1, \gamma_2, \ldots, \gamma_g\rangle$ and Λ is totally disconnected and perfect. So Λ is homeomorphic to a Cantor set.† Also Γ is isomorphic to a free group of rank g. Let $\Omega = \mathbf{S^n} - \Lambda$. Since $\mathbf{S^n} - \bigcup_{i=1}^{g}\{D_i \cup D'_i\}$ is a fundamental domain for Γ one sees that $\Gamma\backslash\Omega$ is homeomorphic to a connected sum of g copies of $\mathbf{S^{n-1}} \times \mathbf{S^1}$. It is called a *Schottky manifold.*

One of Schottky's interests in the construction is the theorem: every compact Riemann surface of genus $g \geq 1$ can be obtained as a Schottky manifold.

(6.8) Connected sums The possibilities of inversions in conformal or Möbius geometry give rise to some very interesting phenomena. Here is one such simple situation. Let M_i^n i=1,2, be two conformally flat (or Möbius) manifolds, and $D_i^n \subseteq$

† This is one of the very early examples of infinite processes in mathematics leading to a Cantor set. With a historical hindsight it seems amazing that Schottky ran into a Cantor set around the same time Cantor was struggling with his set-theoretic foundations of mathematics !

M_i^n i=1,2, be two round n-diska. (If M_i^n is only known tobe conformally flat we choose the round disks insome admissible co-ordinate charts.) Let $M_{oi}^n = M_i^n - (int\, D_i^n)$, so ∂M_{oi}^n is a round $\mathbf{S^{n-1}}$.

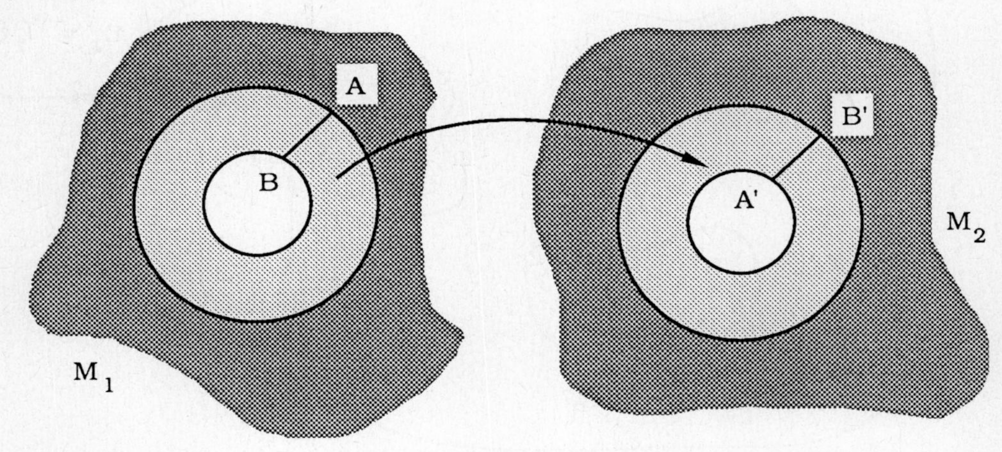

Figure 10

Since one can invert in round (n -1)–spheres, it is clear that one can glue M_{o1}, M_{o2} along their boundaries so that the resulting manifold (which is differentiably an oriented or unoriented connected sum) also admits a conformally flat (or Möbius) structure. Various choices of D_i and various glueing possibliities may result in the same ambient differentiable manifold, but in general will result in mutually inequivalent Möbius structures. In the context of compact Riemann surfaces this construction gives a rough but correct count of Riemann's moduli.

It can be shown that if both M_1^n, M_2^n are Kleinian then their oriented or unoriented connected sum admits a Kleinian structure, cf. [KP].

In the context of Kleinian groups, this construction is closely associated with Klein's "combination theorems" further embellished by Maskit $[M]_2$.

(6.9) Connected sums along hypersurfaces: cf.$[Ku]_1$. Let M_i^n, $i = 1, 2$ be manifolds with boundaries W_i^{n-1} respectively. Suppose there exists a homeomorphism $h : W_1 \to W_2$. Then the manifold obtained by glueing M_1^n to M_2^n via h is called a *a connected sum along h* (or by abuse, along W_1, W_2) and is denoted by $M_1^n \#_h M_2^n$

Now let M_i^n $i = 1,2$ be Möbius manifolds with boundaries W_i^{n-1} resp. Suppose W_i are connected and their tubular neighborhoods are Möbius–diffeomorphic. Let $h : W_1 \longrightarrow W_2$ be a restriction of such a diffeomorphism. Let $\tilde{M}_i^n$ be the universal covers of M_i and $\tilde{W}_i$ some boundary components of $\tilde{M}_i$ lying over W_i. Suppose now that $\tilde{W}_1$ (hence also $\tilde{W}_2$) is Möbius-diffeomorphic to an open subset of a round $\mathbf{S^{n-1}}$. (Actually $\tilde{W}_i$'s need *not* be universal covers. It would suffice if they are Galois but such that h lifts to a map equivariant w.r.t. the deck transformation groups.)

So the inversion in the round $\mathbf{S^{n-1}}$ may be used to turn a tubular neighborhood of $\tilde{W}_1$ – and hence also that of W_2 – "inside out". This intuitive description should suffice to see that $M_1^n \#_h M_2^n$ admits a Möbius structure.

Here are some special cases of interest.

(6.9.1) **Doubling a hyperbolic manifold:** Let $M^n = \Gamma \backslash D^n$ be a complete hyperbolic manifold. Regard D^n as a round disk in S^n , and Λ as in (6.6) so that Γ acts freely and properly discontinuosly on $\mathbf{S^n} - \Lambda$. The manifold $M_1^n = \Gamma \backslash \{\mathbf{S^n} - \Lambda\}$ is essentially a connected sum of M^n with itself along its "ideal boundary". (It may be disconnected if $\Lambda = \mathbf{S^{n-1}}$.)

(6.9.2) **Cuspidal connected sum:** Let $M_i^n = \Gamma_i \backslash \mathbf{D}^n$, $i = 1,2$ be non-compact hyperbolic manifolds of finite volume. Suppose it is possible to cut off an appropriate neighborhood of a cusp of M_i and obtain M_{oi}^n with boundary such that the boundary component is a Bieberbach manifold. Suppose that the boundary-component of M_{o1}^n is a Möbius–diffeomorphic to that of M_{o2}^n, by a diffeomorphism h. Then $M_{o1} \#_h M_{o2}$ admits a Möbius structure. We call it (by abuse) as a *cuspidal connected sum* of M_1 and M_2.

(6.9.3) Let M_i^n, $i = 1,2$ be a compact Riemannian manifold of constant curvature -1,(resp 1) and with non-empty totally geodesic boundary W_i^{n-1} . Suppose there exists an isometry $h : W_1^{n-1} \longrightarrow W_2^{n-2}$. Then $M_1^2 \#_h M_2^n$ admits a Riemannian metric of constant curvature -1,(resp.1).

This follows since in the hyperbolic space or in $\mathbf{S^n}$ there exist inversions in totally geodesic hyper-surfaces which are isometries !

(6.9.4) **A variation of (6.9.3); deformations of hyperbolic manifolds as Möbius manifolds:** Let M_i^n, $i = 1,2$ be a hyperbolic manifold as in (6.9.3). Let W_{oi}^{n-1} be a hypersurface *parallel* to W_i^{n-1} – i.e. a set of points at a constant distance $\epsilon > o$ (sufficiently small) from W_i^{n-1}. It is a simple fact of hyperbolic geometry that W_{oi}^{n-1} is totally umbilic and so if $\tilde{M}_i^n$ denotes the universal cover considered as embedded in $\mathbf{D^n} \subseteq \mathbf{S^n}$ then each component of the inverse image of W_0^{n-1} in $\tilde{M}_i$ is contained in a round (n-1) – sphere. Let M_{0i}^{n-1} be obtained from M_i^n by cutting off a tubular neighborhood of W_i so that the boundary of M_{0i}^n is W_{0i}. Alternately we could also insert a tubular neighborhood of W_i so that the new boundary component is parallel to the old one. Now a tubular neighborhood of W_{0i} is not isometric but is still Möbius-diffeomorphic to a tubular neighborhood of W_i. So carrying out the operation in (6.9.3) on M_{0i}^n would result in a Möbius manifold which is not isometric but is diffeomorphic to a hyperbolic manifold.

(6.9.4.1) We can look at this construction from a slightly different stand-point. Let M^n be a compact hyperbolic manifold (without boundary) containing a totally geodesic hypersurface W^{n-1}. Let M_0^n be obtained by cutting M^n along W – so that M_0^n has two boundary components W_1^n and W_2^n which are mutually isometric. Now either inserting or deleting suitable tubular neighborhoods of W_i^n so that the new boundary components are parallel to the old ones, and then glueing them back we obtain deformations of M^n in the Möbius category.

From a group-theoretic viewpoint this provides certain co-compact subgroups of SO(n,1) which are rigid in SO(n,1) (for $n \geq 3$) but which deform non-trivially in SO(n+1,1). Existence of infinitesimal deformations for certain subgroups of SO(n,1) in SO(n+1,1) is due to J. Lafontaine [L]. Also see Millson [Mi].

In the classical case of n=2 also these deformations are interesting. They provide examples of the so-called *quasi-fuchsian* groups.

(6.10) Finally we mention that recently Gromov and Lawson [GL] have found some surprising examples of Möbius structures on some non-product circle-bundles over compact orientable surfaces of genus ≥ 2.

§7 Development and Holonomy

(7.1.) Let σ_o denote the canonical Möbius structure on $\mathbf{S^n}$. A smooth immersion $f : M^n \to \mathbf{S^n}$ clearly induces a Möbius structure on M^n which we denote by $f^*\sigma_o$. A remarkable consequence of Liouville's theorem is the following partial converse of this statement.

(7.2) Theorem Let (M^n, σ) be a simply connected Möbius manifold. Then there exists a smooth immmersion $\delta : M^n \to \mathbf{S^n}$ such that $\delta^*\sigma_o = \sigma$. Moreover if δ_1, δ_2 are two such smooth immersions then there exists a unique $g \in M(n)$ such that $\delta_1 = g \circ \delta_2$.

Proof Fix a base-point $*$ in M and let $\wp$ be the space of *paths* starting at $*$ i.e. continuous maps $f : [0,1] \to M$ with $f(o) = *$. We euip $\wp$ with the compact-open topology. Fix an admissible co-ordinate neighborhood U_* of $*$ and $\varphi_* : U_* \to \mathbf{S^n}$ a Möbius embedding.

Let U and V be two admissible neighborhoods in M so that $U \cap V$ is nonempty and connected. Let $\varphi_U : U \to \mathbf{S^n}$, $\varphi_V : V \to \mathbf{S^n}$ be the corresponding Möbius embeddings. Then by Liouville's theorem $\varphi_U \circ \varphi_V^{-1}|_{\varphi_V(U \cap V)}$ is the restriction of a unique element in M(n).

Let f : [0, 1] $\to M$ be a path. Partition [0,1] into

$$o = t_0 < t_1 < t_2 \ldots < t_n = 1,$$

so that $f\,([t_{i-1}, t_i])$ lies in an admissible neighborhood U_i, $i = 1, \ldots n$ with $U_1 = U_*$. By shrinking U_i if necessary we may assume that $U_{i-1} \cap U_i$ is connected,. Fix the Möbius embeddings $\varphi_i : U_i \to \mathbf{S^n}$ with $\varphi_1 = \varphi_*$, and let $g_i \in M(n)$ be the elements such that

$$g_i|_{U_{i-1} \cap U_i} = \varphi_{i-1} \circ \varphi_i^{-1}|_{U_{i-1} \cap U_i}, \; i = 2, 3 \ldots n.$$

Then "we develop f into $\mathbf{S^n}$" as follows:

First, $\varphi_1 \circ f$ maps a neighborhood of $[\,t_0,\, t_1]$ into $\mathbf{S^n}$ and by definition of g_2, we see that $\varphi_1 \circ f$ and $g_2 \circ \varphi_2 \circ f$ coincide in a neighborhoodof t_1. Similarly $g_2 \circ g_3 \circ \varphi_3 \circ f$ coincides with $g_2 \circ \varphi_2 \circ f$ in the neighborhood of $t_2 \ldots$ and so on. In this way we obtain a path

(7.2.1) $\quad \Delta_f$: [0,1] $\longrightarrow \mathbf{S^n}$,

beginning at $\varphi_*(o)$. Define

(7.2.2) $\quad \Delta : \wp \longrightarrow \mathbf{S^n}$,

by setting $\Delta(\mathrm{f}) = \Delta_f(1)$. From the uniqueness part in Liouville's theorem it follows that Δ_f and Δ are independent of the choices of the partition and the co-ordinate neighborhoods. Moreover if g is sufficiently near f (in the compact-open topology) and $g(1) = f(1)$ then $\Delta(f) = \Delta(g)$. But we assumed that M is simply connected, so for *any* path g beginning at $*$ and with $f(1) = g(1)$ we have $\Delta(f) = \Delta(g)$. In other words, Δ actually defines

(7.2.3) $\delta : M^n \to \mathbf{S^n}$.

It is plain that δ is a smooth immersion and $\delta^* \sigma_o = \sigma$. We also see that δ is uniquely determined by the initial choice of φ_*.

Now let δ_1, δ_2 be two Möbius maps $M^n \to \mathbf{S^n}$. Shrinking U_* if necessary we may assume that $\delta_1|_{U_*}$ and $\delta_2|_{U_*}$ are Möbius embeddings. So again by Liouville's theorem there exists a unique $g \in M(n)$ such that $\delta_1|_{U_*} = g \circ \delta_2|_{U_*}$. Now both δ_1 and $g \circ \delta_2$ may be regarded as the maps obtained by the above "development process" with the initial choice of $\varphi_* = \delta_1|_{U_*}$. So by the remarks in the previous paragraph $\delta_1 = g \circ \delta_2$. *q.e.d.*

(7.3) A Möbius manifold (M^n, σ) is said to be *developable* if there exists a Möbius map $\delta : M^n \to \mathbf{S^n}$. In this language the above theorem says that a simply connected Möbius manifold is developable. A Möbius map $\delta : M^n \to \mathbf{S^n}$ is called a *development of* M^n *in* $\mathbf{S^n}$. The argument in the above theorem shows that a development is essentially unique — i.e. any two developments differ by a uniquely determined element of $M(n)$.

(7.4) Let (M^n, σ) be a developable Möbius manifold and $\delta : M^n \to \mathbf{S^n}$ a development. Let A(M) denote a full group of Möbius automorphisms of M. For any $u \in$ A(M), clearly $\delta \circ u$ is also a development and so there exists a unique $\rho(u) \in M(n)$ such that

(7.4.1) $\rho(u) \circ \delta = \delta \circ u$.

Proposition The map $\rho : u \mapsto \rho(u)$ defines a δ–equivariant homomorphism $\mathrm{A(M)} \to M(n)$.

Proof The δ-equivariance is just the statement (7.4.1). Let u_1, u_2, $\in$ A(M). Then by definition of ρ

$$(7.4.2) \qquad \begin{cases} \rho(u_1 \circ u_2) \circ \delta = \delta \circ u_1 \circ u_2 \\ \qquad\qquad\qquad = \rho(u_1) \circ \delta \circ u_2 \\ \qquad\qquad\qquad = \rho(u_1) \circ \rho(u_2) \circ \delta. \end{cases}$$

This shows that ρ is a homomorphism. *q.e.d.*

(7.5) Let (M^n, σ) be a developable manifold. Let δ_1, δ_2 be two development maps. So as seen above $\delta_1 = g \circ \delta_2$ for a uniquely determined $g \in M(n)$. Let ρ_1, ρ_2 be the δ_1-, resp. δ_2- equivariant homomorphisms A(M) $\longrightarrow M(n)$ defined by (7.4). Then for any $u \in$ A(M)

$$\rho_1(u) \circ \delta_1 = \rho_1(u) \circ g \circ \delta_2 = \delta_1 \circ u = g \circ \delta_2 \circ u$$

or

$$(g^{-1} \circ \rho_1(u) \circ g) \circ \delta_2 = \delta_2 \circ u = \rho_2(u) \circ \delta_2$$

Since in (7.4.1) $\rho(u)$ is determined uniquely by u and δ it follows that

(7.5.1) $\rho_2(u) = g^{-1} \circ \rho_1(u) \circ g$

In other words,

Proposition Let (M^n, σ) be a developable Möbius manifold. Then there is a canonical homomorphism ρ : A(M) $\longrightarrow M(n)$ which is determined uniquely upto a conjugacy by an element in M(n).

(7.6) Let (M^n, σ) be a (not necessarily developable) Möbius manifold. Let $p : \tilde{M} \longrightarrow M$ be the universal cover of M and $\tilde{\sigma} = p^*\sigma$ the induced Möbius structure on $\tilde{M}$. Let $\tilde{\delta} : \tilde{M} \longrightarrow \mathbf{S^n}$ be a developable map. By abuse of language one sometimes calls $\tilde{\delta}$ as the *the development map* of (M^n, σ). Let $\pi \approx \pi_1(M)$ be the deck-transformation group acting on $\tilde{M}$. Then clearly $\pi \subseteq$ A$(\tilde{M})$. Let ρ : A$(\tilde{M})$ $\longrightarrow M(n)$ be the homomorphism defined above. Then $\rho|_\pi$ is called *the holonomy* (or *the holonomy homomorphism*) of (M^n, σ).

The geometric–algebraic invariants $\tilde{\delta}$ and ρ which are essentially unique in the sense described in (7.2), (7.5) are the basic invariants of a Möbius structure. They indicate for example the possibilities of deformations of a Möbius structure on the same ambient manifold.

(7.7) A criterion for developability

Propostion Let (M^n, σ) be a Möbius manifold with p, π, $\tilde{\delta}$ and ρ as in (7.6). Then

i) M^n is developable *iff* $\rho\,|_\pi$ is trivial.

ii) More generally, let $\kappa = \ker \rho\,|_\pi$ and M_κ = the covering of M corresponding to κ. Then M_κ is the smallest developable covering of M—in the sense that any other developable covering $M_1 \to M$ factors through M_κ.

Proof Suppose M^n is developable and $\delta : M^n \to \mathbf{S^n}$ its development. Then $\delta \circ p$ is a development map of $\tilde{M}$ and so may be taken as $\tilde{\delta}$. So $\tilde{\delta}$ maps every π-orbit in $\tilde{M}$ to a single point. Hence by (7.4.1), $\rho(\pi) = \{e\}$. The converse follows by reversing these steps. This implies *i)*, and *ii)* follows easily from *i)*. *q.e.d*

(7.8) The above proposition explains the distinction between a general Möbius structure, and a Kleinian structure. In the notation of (7.7) let $\delta_\kappa : M_\kappa \to \mathbf{S^n}$ be a development map. It is clear that (M^n, σ) *is Kleinian iff* δ_κ *is an embedding.* A notable necessary condition for this is $\rho(\pi)$ acts freely and properly discontinuously on im $\tilde{\delta}$, and $\tilde{\delta}$ is a covering map, not just a local homeomorphism† If this condition is satisfied then M is a covering of a Kleinian manifold $\rho(\pi)\backslash im\,\tilde{\delta}$.

Of course the condition can fail in many ways giving rise to several constructions of non–kleinian Möbius structures, and it is also of interset to find sufficient conditions ensuring Kleinian (or "close to Kleinian") nature of a Möbius structure.

Gunning [Gu] found a nice criterian for the classical case: if (M^2, σ) is a compact Möbius manifold $\not\approx (\mathbf{S^2}, \sigma_o)$ then $\tilde{\delta}$ is a covering map iff im $\tilde{\delta} \neq \mathbf{S^2}$, cf. also Kra $[K]_1$, $[K]_2$ for another perspective. In [KP] and in the forthcoming work these results are reproved and extended in many ways.

† A local homeomorphism is a covering iff it has a path-lifting property.

§8 Ideal Boundary, Classification of Möbius Structures

(8.1) There are various notions of "ideal boundary" in the classical theory of Riemann surfaces, cf. [C]. However the notion introducted below is new even in this classical context. We attach an ideal boundary to a developable Möbius manifold. The motivation is that the development map should extend continuously to the ideal boundary.

(8.2) Let (M^n, σ) be a developable Möbius manifold and $\delta : M^n \to \mathbf{S^n}$ a development. Let g_o be a Riemannian metric on $\mathbf{S^n}$ with the standard ambient conformal structure. Let $g = \delta^* g_o$ be the induced Riemannian metric on M^n and $\bar{M}$ the Cauchy completion of M *w.r.t.* the metric (i.e., the distance function) defined by g.

Proposition $\bar{M}$ (as a topological space) does not depend on the choices of g_o and δ.

Proof: Let g'_o, and δ' be other choices and $g' = \delta'^* g'_o$. Since $\mathbf{S^n}$ is compact there exists a constant c *s.t.*

$$\frac{1}{c} g \leq g' \leq c\, g.$$

So g, g' define the same Cauchy sequences. *q.e.d.*

(8.3) The above proposition implies that $\bar{M}$ depends only on σ. We call $\bar{M}$ the *Möbius completion* of M The set

(8.3.1) $$\partial_o M = \bar{M} - M$$

is called *the ideal boundary* of (M, σ).

(8.4) Proposition Let (M, σ) be developable Möbius manifold and $\delta : M \to \mathbf{S^n}$ a development. Then δ extends to $\bar{\delta} : \bar{M} \to \mathbf{S^n}$.

Proof Let g_o be a standard Riemannian metric on S^n and $g = \delta^* g_o$. Let $p_o \in \partial_o M$ and $\{p_n\}, n = 1, 2 \ldots$ a Cauchy sequence in M so that $\lim_n p_n = p_o$. There exist rectifiable arcs in M joining p_n to p_{n+1} and of length arbitrarily close to the distance between p_n and p_{n+1}. So it is easy to see that there exists a rectifiable arc $f : [0, 1) \to M$ so that $\lim_{t\to 1} f(t) = p_o$, and the length of f (as a parametrized arc) is finite. This length equals the length of $\delta \circ f$. Since S^n is compact any sequence $\delta \circ f(t_n)$, $t_n \uparrow 1$ has accumulation points. But since the length of $\delta \circ f$ is finite we see that there is only one such accumulation point. In other words $\lim_{t\to 1} \delta \circ f(t)$ exists. This clearly gives a required extension $\bar{\delta} : \bar{M} \to \mathbf{S^n}$ of δ. *q.e.d* .

(8.5) Proposition Let (M_1, σ_1), (M_2, σ_2) be two developable Möbius manifolds and $f : M_1 \to M_2$ a Möbius map. Then f extends to $\bar{f} : \bar{M}_1 \to \bar{M}_2$. Moreover if $\bar{f}(\partial_\circ M_1) \subseteq \partial_\circ M_2$ then f is a covering and in fact $\bar{f}(\partial_\circ M_1) = \partial_\circ M_2$.

Proof Lt $\delta_2 : M_2 \to \mathbf{S^n}$ be a development. Then $\delta_1 = \delta_2 \circ f$ is a development of M_1. Let $g_\circ$ be a standard Riemannian metric on $\mathbf{S}^n$, and $g_i = \delta_i^* g_o$, $i = 1, 2$ Then $f^* g_2 = g_1$ and so f is a local isometry. Now the argument as in (8.4) shows that f extends to $\bar{f} : \bar{M}_1 \to \bar{M}_2$. This proves *i)*.

Now suppose that $\bar{f}(\partial_\circ M_1) \subseteq \partial_\circ M_2$. Let $u_2 : [0,1] \to M_2$ be a rectifiable path, and $u_1 : [0,1) \to M_1$ is a partial lifting — so $f \circ u_1 = u_2$ on [0,1). The lengths of u_i are equal and finite. So $\lim_{t \to 1} u_1(t)$ exists in $\bar{M}_1$. If this limit does not exist in M_1 then we clearly get a contradiction to the hypothesis that $\bar{f}(\partial_\circ M_1) \subseteq \partial_\circ M_2$. So f has a path–lifting property. Hence f is a covering map.

Next let $p_2 \in \partial_\circ M_2$. Let $u_2 : [0,1) \to M_2$ be a rectifiable path so that $\lim_{t \to 1} u_2(t) = p_2$. Let $u_1 : [0,1) \to M_1$ be a lift of u_2. Since u_1, u_2 have the same finite length $\lim_{t \to 1} u_1(t)$ exists in $\bar{M}_1$. If this limit is p_1 then clearly $\bar{f}(p_1) = p_2$. Also p_1 must lie in $\partial_\circ M_1$, for otherwise p_1 lies in M_1 and so p_2 would lie in M_2. So $\bar{f}(\partial_\circ M_1) = \partial_\circ M_2$. *q.e.d.*

(8.6) Proposition Let (M, σ) be a developable Möbius manifold.

i) *If* $\partial_\circ M = \phi$ *then* $(M, \sigma) \approx (\mathbf{S^n}, \sigma_\circ)$.

ii) *If* $\partial_\circ M = \{$*a point*$\}$ *then* $(M, \sigma) \approx (\mathbf{E^n}, \sigma_\circ|_{\mathbf{E^n}})$.

(we consider $\mathbf{E^n} \approx \mathbf{S^n} - \{\infty\}$.)

Proof Let $\delta: M \to \mathbf{S^n}$ be a development. If $\partial_\circ M = \phi$ then clearly M is compact, and δ is a covering map. Hence $(M, \sigma) \approx (\mathbf{S^n}, \sigma_\circ)$.

Now suppose $\partial_\circ M = \{*\}$, and $\bar{\delta}: M \cup \{*\} \to \mathbf{S^n}$ the extension of δ. We may take $\bar{\delta}(*) = \infty$. Let $M_1 = \bar{M} - \bar{\delta}^{-1}(\{\infty\})$. It is clear that $\bar{M}_1 = \bar{M}$, $\delta(M_1) = \mathbf{E}^n$, and $\bar{\delta}$ maps $\partial_\circ M_1$ into $\partial_\circ \mathbf{E}^n$. So by (8.5), $\delta|_{M_1}$ is a covering map. Since $\mathbf{E}^n$ is simply conected, $\delta|_{M_1}$ is a Möbius homeomorphism. Hence $\partial_\circ M_1 = \{$ *a point* $\}$, so we must have $\partial_\circ M_1 = \{*\}$, $M_1 = M$, and so $(M, \sigma) \approx (\mathbf{E}^n, \sigma_\circ|_{\mathbf{E^n}})$. *q.e.d*

(8.7) The above proposition leads to the fruitful classification of Möbius structures. Let (M, σ) be a Möbius manifold and $(\tilde{M}, \tilde{\sigma})$ its universal cover. We say (M, σ) is *elliptic* if $(\tilde{M}, \tilde{\sigma}) \approx (\mathbf{S^n}, \sigma_\circ)$, and *parabolic* if $(\tilde{M}, \tilde{\sigma}) \approx (\mathbf{E}^n, \sigma_\circ|_{\mathbf{E^n}})$. Otherwise we call (M, σ) *hyperbolic*. We saw in (6.2), (6.3) that elliptic (resp. parabolic) Möbius manifolds are precisely the conformal classes of the spherical (resp. Euclidean) space-forms in the sense of Riemannian geometry, and hyperbolic Möbius manifolds of dimension ≥ 2 properly contain the conformal classes of hyperbolic Riemannian space-forms. In this sense this trichotomy of Möbius manifolds generalizes the wellknown trichotomy among the Riemannian manifolds of constant positive, zero, and negative curvature. Also any Möbius structure on an elliptic (resp. hyperbolic) Riemann surface is elliptic (resp. hyperbolic). On the other hand the canonical Möbius structure on $\mathbf{S}^2 - \{2\,points\}$ is hyperbolic, although as a Riemann surface it is usually considered as parabolic. A Kleinian group is non-elementary,(cf. **(6.5)** for this notion), iff each component of the corresponding Möbius manifold is hyperbolic.

(8.8) A useful property of hyperbolic Möbius manifolds is contained in the following

Proposition Let (M^n, σ) be a developable hyperbolic Möbius manifold. Given $p \in M$ there exists a maximal round n-ball containing p.

Proof Let $\delta: M^n \to \mathbf{S^n}$ be a development. Let $\{B_i\}$ $i = 1, 2 \ldots$ be a maximal increasing family of round n-balls containing p. Let $B = U_{i=1}^{\infty} B_i$. It is clear that $\delta|_{B_i}$ is injective for each i, and hence $\delta|_B$ is also injective. An union of an increasing family of round n-balls in $\mathbf{S^n}$ is either again a round n-ball or $\approx \mathbf{E}^n$. So if B is a round n-ball it is clearly a maximal round n-ball containing p. Otherwise $B \approx \mathbf{E}^n$. we show that this cannot happen. Since M is hyperbolic $\partial_\circ M$ contains at least two distinct points say $\{a, b\}$. Also $\partial_\circ B = \{a\ point\}$ and $\partial_\circ B \subseteq \bar{M}$. Say $\partial_\circ B = \{c\}$. Any properly embedded arc in M starting from p must tend towards c. But surely there exist such arcs tending towards either a or b. Since c cannot be both a and b we have the desired contrdiction. *q.e.d.*

(8.9) Let (M, σ) be a developable Möbius manifold. A round n-ball in M admits a hyperbolic Riemannian mertic. The well-known procedure of Kobayashi in the case of complex manifolds in our context leads to a canonical pseudo-metric on (M, σ). Now (8.8) with further work leads to a conclusion that a hyperbolic Möbius manifold actually admits a canonical Riemannian metric with quite remarkable properties, *e.g.* it shows that a hyperbolic Möbius manifold of dimension n admits a canonical stratificaation by hyperbolic Riemannian manifolds of dimensions $\leq n$. Among other things this metric appears to be useful in explaining the geometric underpinnings of the classical theory of Kleinian groups and exending it in all dimensions, cf. [KP].

References

[A] L.Ahlfors, Finitely generated Kleinian groups, Amer. Jour. of Math. 86 (1964), 413-429; ibid. 87(1965),759.

[B] L Bers, Inqualities for finitely generated Kleinian groups, J.Analyse Math. 18(1967), 23-41.

[C] C.Constantinescu and A. Cornea, Ideale Ränder Riemannscher Flächen, Ergebn. Math. 32, Berlin, Springer-Verlag (1963).

[D] P.Dombrowski, 150 years after Gauss' " disquisitiones generales circa superficies curvas",Astérisque 62, Soc. Math. de France (1979).

[FM] M.Feighn, and D.McCullough, Finiteness conditions for 3-manifolds with boundary, (preprint).

[G] F.Gehring, Rings and Quasiconformal Mappings in Space, Trans. Amer. Math. Soc. 103(1962), 353-393.

[Gr] L.Greenberg, On a Theorem of Ahlfors and conjugate subgroups of Kleinian Groups, Amer. Jour. of Math. 89(1967),56-68.

[GL] M.Gromov, and B.Lawson,(preprint).

[Gu] R.C.Gunning, Special co-ordinate coverings of Riemann surfaces, Math. Ann.. 170(1967), 67-86.

$[K]_1$ I.Kra, Deformations of fuchsian groups, Duke Math J. 36(1969), 537-546.

$[K]_2$ I.Kra, Deformations of fuchsian groups II,

$[K]_3$ I.Kra, (preprint).

[KP] R.S.Kulkarni, and U. Pinkall, Uniformizations of geometric structures and applications to conformal geometry, Diffrential Geometry Peñiscola 1985, Lecture Note in Math. 1209, Springer-Verlag (1986), 190-209.

[KS] R.S. Kulkarni,and P. Shalen, On Ahlfors's finiteness theorem, (to appear in the Advances in Math.)

$[Ku]_1$ R.S.Kulkarni, On the principle of uniformization, Jour. of Diff. Geom. 13(1978), 109-138.

$[Ku]_2$ R.S.Kulkarni, Groups with domains of discontinuity, Math. Ann. 237 (1978),

253-272.

[L] J. Lafontaine, Modules de structures conformes plates et cohomologie de groupes discretes, C.R.Acad.Sci.Paris 297(1983), 655-658.

[M]$_1$ B.Maskit, On the classification of Kleinian groups, Acta Math. 135(1975),249-270; ibid.138(1977) 17-42.

[M]$_2$ B.Maskit, On Klein's combination theorem, Trans. Amer. Math. Soc.120(1965) 499-509; ibid. 131(1968) 32-39.

[M]$_3$ B.Maskit, A theorem on planar regular covering surfaces with applications to 3-manifolds, Ann. of Math. 81(1965) 341-355.

[Mc] D.McCullough, Compact submanifolds of 3-manifolds with boundary, Quarterly J. Math. (2) 37 (1986) 299-307.

[Mi] J. Millson, A remark on Raghunathan's vanishing theorem, Topology 24(1985), 495-498.

[Mo] G.D.Mostow, Strong rigidity of locally symmetric spaces, Ann. of Math. studies 78, Princeton University Press(1973).

[S] D. Sullivan, A Finiteness Theorem for Cusps, Acta Math. 147 (1981), 127-144.

Conjugacy Classes in M(n)

*Ravi S. Kulkarni**

Contents

* Supported by an NSF-grant, and Max-Planck-Institut für Mathematik, Bonn

§0 Introduction

(0.1) From a geometric viewpoint, if two elements in a group G are conjugate then in any G-action these two elements act "similarly," if we consider the conjugating transformation in a passive way i.e., a "renaming" of the points in the space. Hence the importance of the classification of the elements or subgroups of a group into conjugacy classes.

(0.2) In this chapter we shall first classify the elements of M(n) up to conjugaacy. In this connection it is convenient to consider M(n) either as a group of conformal transformations of $\mathbf{S^n}$, or as a group of isometries of the hyperbolic (n+1) - dimensional space in its disk model $\mathbf{D^{n+1}}$, or in the upper half space model $\mathbf{H^{n+1}}$, or in its linear model cf. chapter 1, §4. Also to get a better perspective we separately classify the conjugacy classes in the groups of spherical and Euclidean isometries and the group of Euclidean conformal (or similarity) transformations.

(0.3) In §4 we prove a theorem of Greenberg to the effect that the maximal connected subgroups of M(n) are in some sense the obvious ones admitting a simple geometric description. This is followed in §5 by a proof of extensions of theorems of Nielsen and Van Vleck. The original theorem of Nielsen,cf. [5], says that a non-abelian subgroup of M(1) which consists only of hyperbolic elements besides the identity is necessarily discrete. Similarly the original theorem of Van Vleck, cf.[8], says that a subgroup of M(2), considered as PSL(2,C), which besides the identity consists of hyperbolic elements with real traces necessarily leaves a round circle invariant, i.e. it must be conjugate to a subgroup of M(1), considered as PSL(2,R) $\leq$ PSL(2,C). Greenberg's and Chen-Greenberg's proofs of various extensions of these theorems, cf. [1],[2], use the important geometric notion of a limit set of a subgroup and also some Lie theory. Perhaps one would have liked elementary geometric proofs of these theorems. In Nielsen's and Van Vleck's cases one can make some clever computations involving 2×2 matrices and elementary geometric arguments. However, in general some use of the theories of Lie groups and algebraic groups seems unavoidable, and perhaps there is not much point in avoiding it. In our proofs we have freely appealed to some of the standard structure- and classification- parts of these theories. These proofs are susceptible to further generalizations.

(0.4) In §6 we prove that unless a subgroup of M(n) has a fixed point in $\mathbf{D^{n+1}}$ or $\mathbf{S^n}$ it must contain a hyperbolic element. On the other hand if a subgroup of M(n) contains no hyperbolic element then it must have a fixed point in $\mathbf{D^{n+1}}$ or $\mathbf{S^n}$. For $n \leq 2$ these statements occur implicitly in the classical literature and can be proved again by some clever computations in 2×2 matrices and appeals to geometry. The proofs of these results valid in all dimensions appear to be new.

(0.5) In §7 we prove a theorem on subgroups of M(n) consisting of elliptics only. It is curious that for $n \leq 3$ such a subgroup must be conjugate to a subgoup of O(n + 1), but for $n \geq 4$ this is no longer so. This result does not seem to have appeared in print. But the existence of subgroups consisting of elliptics without a fixed point in $\mathbf{D^n}$ for $n \geq 4$ was pointed out to me by P. Waterman. These considerations come up naturally if one wishes to develop a theory of limit sets for not necessarily discrete groups.

§1. The Conjugacy Classes in E(n) and Sim(n).

(1.1) Let V $\approx R^n$ be the real vector space with the positive definite inner product, and O(V) = O(n) the corresponding group of orthogonal transformations. Let g be in O(V). As is well known, $V = \oplus V_i$ where V_i are 1 - or 2-dimensional g–invariant irreducible subspaces. If $dim\ V_i = 1$ then g $|_{V_i}$ acts as [1] or [-1]. We then assign to $g\ |_{V_i}$ the *rotation angle* 0 or π respectively. If $dim\ V_i = 2$ then g $|_{V_i}$ is a rotation through an angle $\theta, 0 < \theta < 2\pi$, $\theta \neq \pi$. We then assign the *rotation angle* θ to g $|_{V_i}$. It is well-known that

Proposition The rotation angles form a complete set of invariants of a conjugacy class in O(V).

(1.2) Now consider the group E(n) of Euclidean motions of $\mathbf{E^n}$. For g in E(n) set

(1.2.1) $$\lambda(g) = \inf_{\vec{x} \in \mathbf{E^n}} ||g\vec{x} - \vec{x}||.$$

Then λ(g) is called the *translational length* of g. If g is a *translation*, i.e. of the form $\vec{x} \mapsto \vec{x} + \vec{a}$ then $\lambda(g) = ||a||$. If g has a fixed point then g is called a *rotation*. If g is a rotation then λ(g) = o.

(1.3) Since $E(n) \approx R^n \lhd O(n)$ there is a homomorphism

(1.3.1) $$\rho : E(n) \rightarrow O(n)(\approx E(n)/R^n).$$

(The last isomorphism depends on the choice of the origin. But one easily checks that ρ is independent of this choice.) The rotation angles of an element g in E(n) are by definition the rotation angles of ρ(g) in the sense of (1.2). Clearly if g is a translation then its rotation angles are all zero.

(1.4) There is a third kind of Euclidean motion which for want of a better name we shall call a *transrotation*. Let V_1 be a k-dimensional subspace of $E^n, 0 < k < n$. Choosing the origin in V_1 we can express $E^n = V_1 \oplus V_2$ as a sum of orthogonal linear subspaces. Consider a motion which leaves V_1 invariant and acts as a translation in V_1 and rotates around V_1 having no set of oriented parallel lines orthogonal to V_1 invariant. In terms of appropriate coordinates in V_1 and V_2 such a motion has the form

$$(1.4.1) \qquad \begin{bmatrix}\vec{x}_1\\ \vec{x}_2\end{bmatrix} \longmapsto \begin{bmatrix} I & 0\\ 0 & A_2\end{bmatrix}\begin{bmatrix}\vec{x}_1\\ \vec{x}_2\end{bmatrix} + \begin{bmatrix}\vec{b}_1\\ \vec{0}\end{bmatrix},\ \vec{b}_1 \neq \vec{0},$$

where A_2 in $O(V_2)$ has no rotation angle equal to zero. Such a motion will be called a *transrotation*. For a transrotation in the form (1.4.1), the translational length is $||\vec{b}_1||$ and the rotation angles are those of A_2 and the angle 0 which occurs $k = dim\, V_1$ times.

If n = 2, k = 1 then a transrotation is also called a *glide reflection*. It is necessarily orientation-reversing. If n = 3 then a transrotation with k = 1 is called a *screw − motion*.

(1.5) Theorem The translational length and rotation angles form a complete set of invariants of a conjugacy class in E(n). Moreover

(1.5.1) λ(g) = o iff g is a rotation.

(1.5.2) All rotation angles are zero iff g is a translation.

(1.5.3) λ(g) $\neq$ o and some rotation angle $\neq$ o iff g is a transrotation.

$$(1.5.4) \qquad \lambda(g) = \min_{\vec{x} \in E^n} ||g\vec{x} - \vec{x}||$$

In particular every Euclidean motion is either a translation, or a rotation, or a transrotation.

Proof (a sketch) An Euclidean motion with respect to an orthonormal coordinate system can be expressed in the form:

$$(1.5.5) \qquad g : \vec{x} \mapsto A\vec{x} + \vec{a}, A \in O(n), \vec{a} \in R^n.$$

Then ρ(g) = A and the rotation angles of A are those of g. If all rotation angles of g are zero then A = I and g is a translation. If no rotation angle of g is zero then A does not have 1 as an eigenvalue. So the equation

$$A\vec{x} + \vec{a} = \vec{x}$$

has a solution. So g has a fixed point and hence g is a rotation. Now suppose exactly k rotation angles of A are zero and $o < k < n$. Let U_1 be the eigenspace of A with eigenvalue 1 , and U_2 its orthogonal complement. Write a point in E^n in the form $\vec{x}_1 + \vec{x}_2, \vec{x}_i \in U_i$, i= 1, 2. So g has the form

$$\begin{bmatrix}\vec{x}_1\\ \vec{x}_2\end{bmatrix} \longmapsto \begin{bmatrix} I & 0\\ 0 & A_2\end{bmatrix}\begin{bmatrix}\vec{x}_1\\ \vec{x}_2\end{bmatrix} + \begin{bmatrix}\vec{b}_1\\ \vec{b}_2\end{bmatrix}$$

Now A_2 has no eigenvalue equal to 1 so

$$A_2\vec{x}_2 + \vec{b}_2 = \vec{x}_2$$

has a solution, say $\vec{x}_2^\circ$. Translate the coordinate system to $\begin{bmatrix} 0 \\ \vec{x}_2^\circ \end{bmatrix}$ and still call the new coordinates $(\vec{x}_1, \vec{x}_2)$. So g now has the form

$$\begin{bmatrix} \vec{x}_1 \\ \vec{x}_2 \end{bmatrix} \longmapsto \begin{bmatrix} I & 0 \\ 0 & A_2 \end{bmatrix} \begin{bmatrix} \vec{x}_1 \\ \vec{x}_2 \end{bmatrix} + \begin{bmatrix} \vec{b}_1 \\ \vec{0} \end{bmatrix}.$$

If $\vec{b}_1 = \vec{o}$ then g is still a rotation; otherwise it is a transrotation. We have thus shown that every Euclidean motion is either a translation, a rotation, or a transrotation. In each case it is easy to see that $\lambda(g)$ is actually the min $||g\vec{x} - \vec{x}||$. So we get (1.5.4). The other assertions are fairly clear and are left to the reader. q.e.d.

(1.6) We remark that the rotation-angles (counted mod 2π) are continuous functions on E(n) , but the *translational length is not*!

(1.7) From the view-point of its dynamics on $\mathbf{E^n}$ a transrotation is closer to a translation than a rotation. So it is not unreasonable to group the translations and transrotations together and call them *parabolics*. The rotations will be called *elliptics*.

(1.8) Relative sizes of elliptics and parabolics. E(n) has two components: $E_+(n)$ and $E_-(n)$ consisting of orientation-preserving (resp. orientation-reversing) Euclidean motions.

Proposition If n is even then the elliptics form a dense set with non-empty interior in $E_+(n)$, whereas the parabolics form a dense set with non-empty interior in $E_-(n)$. 2) If n is odd, then the parabolics form a dense set with non-empty interior in $E_+(n)$, whereas the elliptics form a dense set with non-empty interior in $E_-(n)$.

Proof Consider the homomorphism

$$\rho : E(n) \to O(n),$$

cf. (1.3). Let $O_+(n)$, resp. $O_-(n)$ be the components of O(n) consisting of the orientation-preserving, resp. orientation-reversing rotations. If n is even the subset S of $O_+(n)$ consisting of rotations with no rotation-angle equal to zero is an open, dense subset of $O_+(n)$. Clearly $\rho^{-1}(S)$ is an open, dense subset of elliptics in $O_+(n)$. The proofs of the other assertions are similar. q.e.d.

(1.9) Now consider the group Sim(n) of conformal, (or similarity) transformations of E^n, cf. chapter 1, (4.2). We have a canonical homomorphism

$$\tau : Sim(n) \to O(n) \times R_+$$

which extends ρ in (1.3). If $\tau(g) = (A \ , \mu)$ the rotation-angles of A are called the *rotation – angles* of g, and μ is called the *multiplier* of g. We shall call g *hyperbolic* if its multiplier is different from 1. An elliptic or parabolic element in E(n) is again called elliptic or parabolic in Sim (n) as well. A hyperbolic element has a unique fixed point in $\mathbf{E^n}$. An element in Sim(n) is parabolic iff it has no fixed point in $\mathbf{E^n}$.

Clearly the set of hyperbolic elements is open and dense in Sim(n). In the context of Sim(n) the translational length as defined in (1.2.1) has no meaning. We set

(1.9.1) $$\lambda(g) = \begin{cases} 0 & \text{if g is elliptic or hyperbolic,} \\ 1 & \text{if g is parabolic.} \end{cases} .$$

The proof of the following proposition may be left to the reader.

Proposition The rotation angles, the multiplier, and the λ-invariant as defined in (1.9.1) form a complete set of invariants of a conjugacy class in Sim(n).

(1.10) A picture for $\mathbf{E_+(2)}$: As a topological space the group $E_+(2)$ is homeomorphic to $\mathbf{S^1} \times \mathbf{R^2}$ i.e., an open solid torus. The subgroup

(1.10.1)
$$\vec{x} \mapsto A\vec{x}, A \in SO(2), \vec{x} \in R^2$$

carries the fundamental group. A parabolic element in this case is a translation. Together with identity, the parabolic elements form a closed subgroup $\approx R^2$.

The justification of the following picture may be left to the reader.

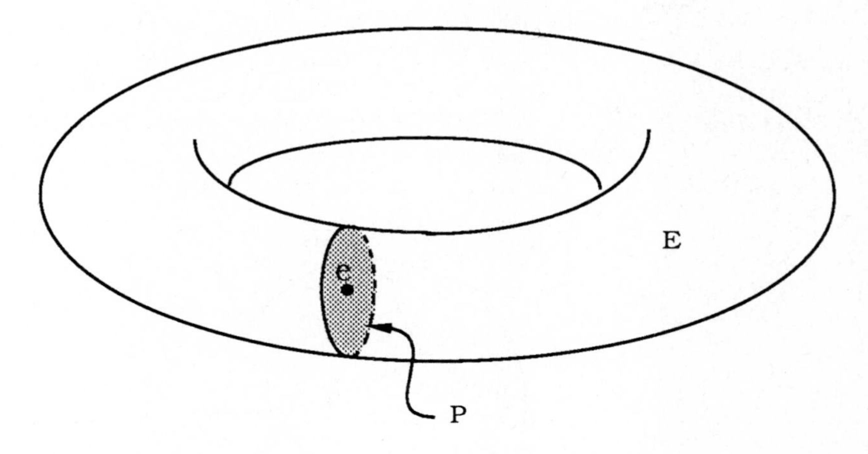

Figure 1.

§2. The Conjugacy Classes in M(n)

(2.1) Proposition Consider M(n) as acting on $\mathbf{D^{n+1}} \cup \mathbf{S^n} = \bar{\mathbf{D}}^{n+1}, n \geq 1$. Let $g \in M(n)$. Then

i) g has at least one fixed point in $\bar{\mathbf{D}}^{n+1}$.

ii) g has either a fixed point in $\mathbf{D^{n+1}}$ or has at most two fixed points in $\mathbf{S^n}$.

Proof i) is immediate from the Brouwer's fixed point theorem. Now suppose g has three distinct fixed points a, b, c in $\mathbf{S^n}$.

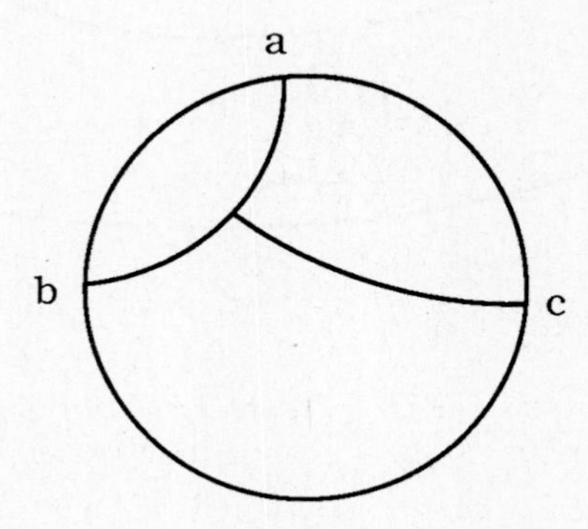

Figure 2.

Then g fixes the foot of the perpendicular from c to the unique geodesic joining a and b. So g has a fixed point in $\mathbf{D^{n+1}}$. This proves ii). q.e.d.

(2.2) An element g in M(n) is called *elliptic* if it has a fixed point in $\mathbf{D^{n+1}}$. A non-elliptic g in M(n) is called *parabolic* if it has a single fixed point in $\mathbf{S^n}$, and *hyperbolic* if it has two distinct fixed points in $\mathbf{S^n}$. In view of (2.1) every element of M(n) is either elliptic, or parabolic, or hyperbolic, and these cases are mutually exclusive.

(2.3) Let g be in M(n). First suppose g is elliptic and p$\in$ $\mathbf{D^{n+1}}$ its fixed point. Then the rotation angles of the differential dg on the tangent space $T_p(\mathbf{D^{n+1}})$ are called the rotation angles of g. (Alternately we may use the linear model for M(n), cf. chapter 1 §4, and conjugate g in O(n+1) $\leq$ M(n) and read the rotation angles there. This shows that the rotation angles do not depend on the choice of p. One may derive these facts also from the hyperbolic geometry of $\mathbf{D^{n+1}}$.) Next suppose g is parabolic, and p its unique fixed point in $\mathbf{S^n}$. The induced action of dg on $T_p(\mathbf{S^n})$ is orthogonal and has 1 as an eigenvalue, and has determinant 1. The rotation-angles of dg_p are called the $rotation - angles$ of g. (Alternately, treating p as ∞ we may consider g as a parabolic element of E(n) and read the rotation-angles there.) Finally suppose g is hyperbolic, and p,q its fixed points in $\mathbf{S^n}$. Let u be the hyperbolic geodesic joining p to q. So g induces an isometry of u $\approx$ $\mathbf{E^1}$, which must be a translation since g has no fixed point in $\mathbf{D^{n+1}}$. Suppose for definiteness that g is a translation on u in the direction of q. Then q is called *the attracting fixed point*, and correspondingly p *the repelling fixed point* of g. Also dg_p is a homothety i.e. conjugate to an element of $O(n) \times R_+$. This allows us to assign the rotation-angles and the multiplier $\geq$ 1 to dg_p. We define the rotation-angles and the multiplier of g to be those of dg_p. (Alternately treating q as ∞ as we may consider g as a hyperbolic element of Sim(n), and read its rotation-angles and multiplier there.)

It is convenient to define the multiplier of an elliptic or parobolic to be 1, and define the λ-invariant (cf. 1.9.1) to be o for elliptics and hyperbolics, and 1 for parabolics.

(2.4) Proposition The rotation-angles, the multiplier †, and the λ-invariant as defined in (2.3) form a complete set of invariants of conjugacy class in M(n).

Proof Notice that we also need to check that the rotation-angles – etc. are indeed invariant under conjugation. An elliptic element in M(n) is conjugate to an element in O(n+1) and two elements in O(n+1) are conjugate in O(n+1) iff they are conjugate in M(n) – this follows by the standard Lie theory since O(n+1) is a maximal compact subgroup of M(n), or alternately by appealing to the linear model in chapter 1,§4, or also the hyperbolic geometry in $\mathbf{D^{n+1}}$. This shows that the rotation-angles are conjugacy-invariants for elliptics. Moreover $\lambda = 0, \mu = 1$ determines the ellipticity of an element. So the proposition is proved for elliptic elements.

Next let g and h be two conjugate parabolic elements. So there exists k $\in$ M(n) with $h = kgk^{-1}$. If p is the fixed point of g, then kp is the fixed point of h. Now

$$dk_p : T_p(\mathbf{S^n}) \rightarrow T_{kp}(\mathbf{S^n})$$

† In contrast with Sim(n), we have multipliers always $\geq$ 1 in M(n).

is a homothety. This easily implies that the rotation-angles of a parabolic are conjugacy-invariants. On the other hand, $\lambda = 1$, $\mu = 1$ determines the parabolicity of an element. This fact, and the transitivity of M(n) on $\mathbf{S^n}$ proves the proposition for parabolic elements.

The proof for the hyperbolic elements is similar and may be left to the reader. q.e.d.

(2.5) Relative sizes of elliptics, parabolics and hyperbolics. The group M(n) has two components $M_+(n)$ and $M_-(n)$ consisting of the orientation-preserving and orientation-reversing Möbius transformations respectively. Let E, P, H denote the sets of elliptics, parabolics and hyperbolics in M(n) respectively.

Proposition 1) Both $H \cap M_+(n)$ and $H \cap M_-(n)$ are open and non-empty. Moreover $H \cap M_+(n)$ is dense in $M_+(n)$ iff n is even, and $H \cap M_-(n)$ is dense in $M_-(n)$ iff n is odd.

2) $E \cap M_+(n)$ has non-empty interior iff n is odd, whereas $E \cap M_-(n)$ has non-empty interior iff n is even.

3) In any case $H \cup E$ contains an open and dense subset of M(n).

Proof The map

(2.5.1)
$$g \in P \mapsto the\, fixed\, point\, of\, g\, in\, \mathbf{S^n}$$

makes P the total space of a fiber bundle with base $\mathbf{S^n}$ and fiber $\approx$ the set of parabolics in Sim(n), or what is the same, in E(n). So

(2.5.2)
$$dim\, P = dim\, \mathbf{S^n} + dim\, E(n) < dim\, M(n).$$

Similarly the map

(2.5.3)
$$g \in H \mapsto the\, attracting\, fixed\, point\, of\, g\, in\, \mathbf{S^n}$$

makes H a fiber bundle with base $\mathbf{S^n}$ and fiber $\approx$ the set of hyperbolics in Sim(n). The latter is open and dense in Sim(n). So

(2.5.4)
$$dim\, H = dim\, \mathbf{S^n} + dim\, Sim(n) = dim\, M(n).$$

The topology of H as a fiber-bundle is the manifold-topology, so it coincides with the subspace-topology induced from M(n). So H is an open subset of M(n). It is easy to see that both $H \cap M_+(n)$ and $H \cap M_-(n)$ are non-empty.

As for elliptics, for definiteness, we restrict to the component $M_+(n)$. Consider the case n odd. Let

(2.5.5)

$$E'_+(n) = \{g \in E \cap M_+(n) \mid g\, has\, no\, rotation\, angle \;=\; 0\}, \quad (n\, odd).$$

Then each g in $E'_+(n)$ has a unique fixed point in $\mathbf{D^{n+1}}$. The map

(2.5.6)

$$g \in E'_+(n) \mapsto the\, fixed\, point\, of\, g\, in\, \mathbf{D^{n+1}}$$

makes $E'_+(n)$ a fiber-bundle with base $\mathbf{D^{n+1}}$ and fiber $\approx$ an open, dense subset in $O_+(n)$. So

(2.5.7)

$$dim\, E'_+(n) = dim\, \mathbf{D^{n+1}} \;+\; dim\, O_+(n) \;=\; dim\, M(n)$$

As in the case of H, we conclude that $E'_+(n)$ is an open subset of M(n), for n odd. Now consider the case n even. Let

(2.5.8)

$$E'_+(n) = \{g \;\in\; E \;\cap\; M_+(n) \;\mid\; g\, has\, exactly\, one\, rotation\, angle \;=\; 0\}.$$

Each g in $E'_+(n)$ has exactly two fixed points in $\mathbf{S^n}$. Let

(2.5.9)

$$B = \{(\mathbf{S^n} \times \mathbf{S^n}) - \Delta\}/\sim$$

where Δ is the diagonal and $(x,y) \sim (y,x),\; x,y \in \mathbf{S^n}$. The map

(2.5.10)

$$g \in E'_+(n) \mapsto (the\, unordered\, pair\, of\, fixed\, points\, of\, g\, in\, \mathbf{S^n}) \;\in\; B$$

makes $E'_+(n)$ a fibration with base B and fiber $\approx$ an open dense subset of $O_+(n)$. So

(2.5.11)

$$dim\, E'_+(n) \;=\; dim\, B \;+\; dim\, O_+(n) \;<\; dim\, M(n).$$

The subset $E_+(n) - E'_+(n)$ may be similarly decomposed into finitely many strata of still smaller dimensions. So in view of (2.5.2) and the fact that H is open, we see that when n is odd $E_+(n) \cup P$ is a closed subset with empty interior.

This proves the parts of 1) and 2) concerning $M_+(n)$. The parts concerning $M_-(n)$ are proved similarly. q.e.d.

(2.6) A picture of $\mathbf{M_+(1)}$, cf.[4], §3. The group $M_+(1) \approx PSL(2,R)$ is homeomorphic to $\mathbf{S^1} \times \mathbf{R^2}$ i.e. an open solid torus. We denote the image of L= $\left[\begin{smallmatrix} ab \\ cd \end{smallmatrix}\right]$, ad - bc = 1 in $PSL_2(R)$ by $L^- = \left[\begin{smallmatrix} ab \\ cd \end{smallmatrix}\right]^-$.

(2.6.0)

$$G = PSL(2,R) \approx M_+(1).$$

(2.6.1)

$$K = \left\langle k_\theta = \begin{bmatrix} c\,s \\ -s\,c \end{bmatrix}^- \middle| c = cos\theta, s = sin\theta, 0 \leq \theta \leq \pi \right\rangle.$$

(2.6.2)

$$A = \left\langle a_\lambda = \begin{bmatrix} \lambda\,0 \\ 0\,\lambda^{-1} \end{bmatrix}^- \middle| \lambda \in R_+ \right\rangle.$$

(2.6.3)

$$N = \left\langle n_\mu = \begin{bmatrix} 1\mu \\ 01 \end{bmatrix}^- \middle| \mu \in R \right\rangle$$

Clearly K $\approx$ SO(2) carries the fundamental group $\approx \mathbf{Z}$ of G. The conjugacy class of $k_\theta, 0 < \theta < \pi$ in G is homeomorphic to G/K which in turn is homeomorphic to an open 2-disk. On the other hand, each conjugacy class of $a_\lambda, \lambda > 0, \lambda \neq 1$ (resp. $n_\mu, \mu \neq 0$) in G is homeomorphic to G/A (resp. G/N), each of which, in turn, is homeomorphic to an annulus.

As in (2.5), let E, P, H denote the sets of elliptics, parabolics and hyperbolics respectively. For g in G let C(g) denote the conjugacy class of g in G. Then

(2.6.4) $\quad E = \{e\} \cup \bigcup_{0<\theta<\pi} C(k_\theta)$,

(2.6.5) $\quad P = C(n_1) \cup C(n_{-1})$,

(2.6.6) $\quad H = \bigcup_{\lambda>1} C(a_\lambda)$

Notice that $C(a_\lambda) = C(a_{\lambda^{-1}}), \lambda > 1$, and it has the multiplier λ^2. Also each of $C(k_\theta)$, and, $C(a_\lambda)$ are closed subsets of G. A careful analysis via the polar decomposition of matrices leads to the following picture of $M_+(1)$.

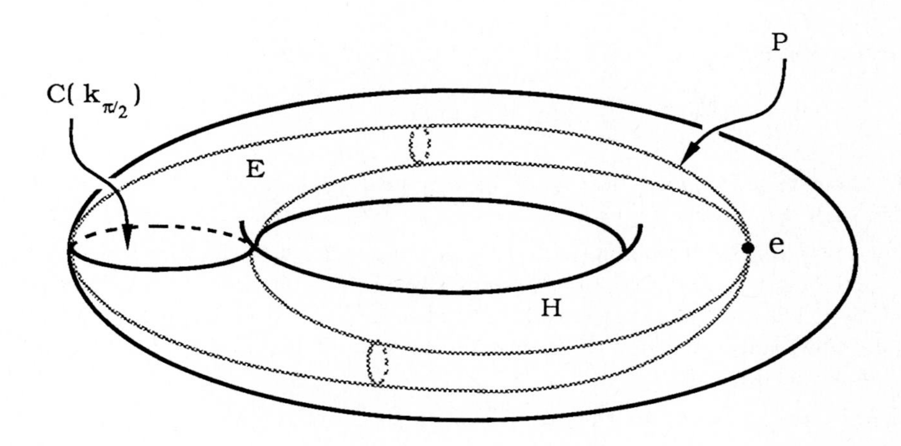

Figure 3

Here $P \cup \{e\}$ is a homeomorphic to a right circular cone $(x^2 + y^2 - z^2 = 0)$ dividing G into two components. One of the components (the "interior" of the cone) is $E - \{e\} \approx \mathbf{R}^3$. The other component (the "exterior" of the cone) is $H \approx \mathbf{S^1} \times \mathbf{R^2}$ One of the $C(k_\theta)$'s, $0 < \theta < \pi$, which for convenience we have chosen to be $C(k_{\pi/2})$ in the picture, may be represented by a "meridinal disk" which divides $E - \{e\}$ into two components. In one of the components $C(k_\theta)$'s, $0 < \theta < \pi/2$, are situated in a snake-like fashion tending towards the "ideal boundary" of $C(k_{\pi/2})$. The same remark applies to the other component which is filled by $C(k_\theta)$'s, $\pi/2 < \theta < \pi$.

The two ends of each $C(a_\lambda), \lambda > 1$ also tend towards the "ideal boundary" of $C(k_{\pi/2})$.

For more details on this picture, and the natural ambient Lorentzian geometry of constant curvature 1 we refer to [4], §3. This space has a natural conformal Lorentzian ideal boundary which is homeomorphic to a 2-torus.

§3 Hyperbolic Translational Length

(3.1) Let d(x,y) denote the hyperbolic distance in $\mathbf{D}^{n+1}$. For g in M(n) set

(3.1.1) $\nu(g) = \inf_{x \in \mathbf{D}^{n+1}} d(x, gx)$.

It is called the *hyperbolic translational length*, or *h – length* for short, of g. Compare (1.2.1).

(3.2) Proposition Let g be in M(n), and μ(g) its multiplier. Then

$$\nu(g) = ln\, \mu(g).$$

Proof. Case 1: *g is elliptic.* If x is the fixed point of g then $d(x, gx) = 0$. So ν(g) = 0 Also μ(g) =1. So ln μ(g) = 0.

Case 2: *g is parabolic.* Let p be the fixed point of g in $\mathbf{S}^n$. Considering p as ∞ and $\mathbf{S}^n - \infty$ as $\mathbf{E}^n$ we know from §1 that g leaves a line in $\mathbf{E}^n$ invariant along which it is a translation. This means that g leaves a round circle, say C_o, through ∞ invariant. Let C_o be the boundary of the 2-dimensional hyperbolic plane $D_o \subset \mathbf{D}^{n+1}$, which is also left invariant by g.

Consider D_o as the upper half-plane. Then up to conjugacy g (restricted to this upper half plane) may be taken to be $z \mapsto z + 1$ where $z = x + iy$ is the complex co-ordinate. It follows that

$$d(x_0 + iy_0, x_0 + 1 + y_0) = \int_0^1 \frac{dx}{y_0} = \frac{1}{y_0}.$$

So as $y_0 \to \infty$ we see that $d(x_0 + iy_0,\, x_0 + 1 + iy_0) \to 0$. So $\nu(g) = 0$.† But $\mu(g) = 1$. So ln μ(g) = 0 also.

---

† Notice that $\nu(g) = 0$ but the d(x, gx) does not attain its minimum for any x in $\mathbf{D}^{n+1}$.

Case 3: *g is hyperbolic.*: Let p, q be the fixed points of g, and $c_\circ$ the the hyperbolic geodesic joining p to q.

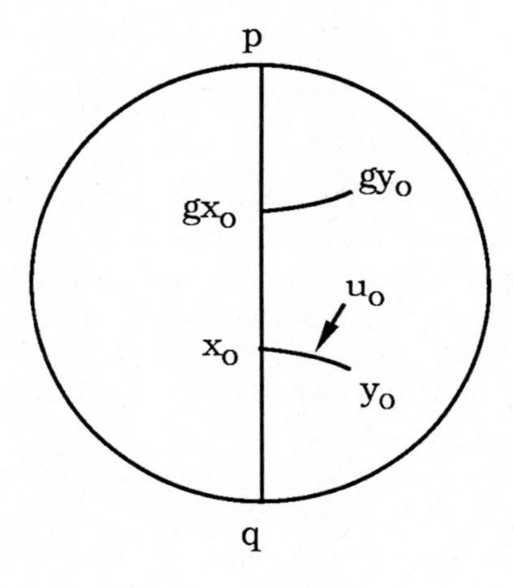

Figure 4.

Then g translates $c_\circ$ say towards q by a length l.. So $d(x, gx) = l$ for all $x \in c_\circ$. Let $y_\circ$ be any point in $\mathbf{D^{n+1}}$ and $u_\circ$ the perpendicular from $y_\circ$ on $c_\circ$ with foot $x_\circ$. Then $gx_\circ$ is the foot of the perpendicular $gu_\circ$ from $gy_\circ$ on c. Let $z_\circ$ be the foot of the perpendicular from $y_\circ$ on the geodesic containing $gu_\circ$. Since $c_\circ$ is the common perpendicular to $u_\circ$ and $gu_\circ$ we see that

$$d(x_\circ, gx_\circ) \leq d(y_\circ, z_\circ) \leq d(y_\circ, gy_\circ)$$

It follows that $\nu(g) = l$. On the other hand treating q as ∞ and p as 0 in $\mathbf{E^n}$ in the upper-half space model $\mathbf{H^{n+1}}$, g is given by

$$\vec{x} \mapsto \mu \begin{bmatrix} A & 0 \\ 0 & 1 \end{bmatrix} \vec{x}, \; \vec{x} \in \mathbf{H^{n+1}}, \; A \in O(n), \; \mu \in R_+.$$

So

$$l = \int_1^{\mu} \frac{dt}{t} = ln\,\mu$$

This proves the proposition for hyperbolic elements. q.e.d.

§4 A Theorem of Greenberg

(4.1) Geometric Subgroups: Let G($\mathbf{S^k}$) denote the full subgroup of M(n) which leaves a round k-sphere $\mathbf{S^k}$ invariant. When the reference to a particular $\mathbf{S^k}$ is not important we shall also denote G($\mathbf{S^k}$) simply by G(k). For k = 0 we consider a round 0-sphere as a pair of points, and for $k = -1$ we consider a (-1)-sphere as a point. So $G(0) \approx O(n) \times R_+$, and $G(-1) \approx Sim(n)$. Finally we set G(-2) to be O(n+1) - which, of course, is defined only upto conjugacy. Let $G_\circ(k)$ denote the identity component of $G(k), k = -2, -1, 0, \ldots n$. Note that G(-2) and G(n) have two components and the other G(k) have four components. Also the conjugacy class of G(k) depends only on k.These G(k)'s, and their subgroups containing $G_\circ(k)'s$ will be called the *geometric subgroups* of M(n). Moreover there are no mutual inclusions among $G(k)$'s for different k's, and also no inclusions among different $G_\circ(k)$'s for different k's except for $G_\circ(0) \subset G_\circ(-1)$. The latter inclusion, of course, is also defined only upto conjugacy. The following interesting theorem about these subgroups is due to Greenberg, cf.[2].

(4.1.1) **Theorem** Maximal proper connected subgroups of M(n) are precisely the conjugacy classes of $G_\circ(k)'s,\ k \neq 0$, or n.

Note that a maximal connected subgroup is closed, so it is a Lie subgroup of M(n).

The proof extends over the rest of the section.

(4.2) Let G be a connected Lie subgroup of M(n). We consider the action of M(n) on $\bar{\mathbf{D}}^{\mathbf{n+1}}$, as well as its standard linear action, cf. chapter 1, §4. As is well-known (and easy to see) the components of the fixed point set of a group of isometries in a Riemannian manifold are totally geodesic submanifolds. Since any two points in $\mathbf{D^{n+1}}$ can be connected by a unique geodesic it also follows that the fixed point set of a group of hyperbolic isometries is connected, so such a set is of the form $\mathbf{D^k}$ - a k-dimensional hyperplane in $\mathbf{D^{n+1}}$. For k = 0, $\mathbf{D^0}$ is a point. We shall also often use the following simple fact: in any action of a group G on a space X the fixed point set of a subgroup H in X is invariant under the normalizer of H in G.

(4.3) Assertion 1: If G contains a non-trivial compact normal subgroup H then G is contained in some $G_o(k)$, $k < n$.

Proof Since H is conjugate to a subgroup of O(n+1), it has a fixed point in $\mathbf{D}^{n+1}$. Let $\mathbf{D}^{k+1}$, $k \geq -1$, be the full fixed point set of H. If $G_o(k)$ is the full subgroup of M(n) leaving $\mathbf{D}^{k+1}$ invariant, it is clear from the remark in (4.2) that $G_o \leq G_o(k)$. q.e.d.

(4.4) Assertion 2: If G contains a non-trivial connected abelian normal subgroup H then G_o is contained in some $G_o(k)$, $k \neq n$.

Proof Clearly we may assume that H is a closed subgroup. In view of (4.3) we may also assume that it is noncompact. So H contains a closed subgroup $\approx R$. Since the closure of the group generated by an elliptic is compact,it follows that H contains a non-elliptic element h. The fixed point set of h in $\mathbf{S}^n$ consists of one or two points, and it is left invariant by H (since H normalizes $< h >$.) Since H is connected in fact H leaves each of these points fixed. But then since H is normal in G and G_o is connected the fixed points of H are fixed by G_o as well. So $G_o \leq G_o(-1)$. q.e.d.

(4.5) Let $G_o = R.S$ be the Levi decomposition, where R is the radical and S a semisimple subgroup of G. Then R is connected. Suppose R is non-trivial and H the last non-trival subgroup in the derived series of R. Then H is a non-trivial connected abelian normal subgroup of G. In view of (4.3) and (4.4) it follows that

Assertion 3: To prove the theorem we may assume that G is semisimple, and non-compact.

(4.6) For any Lie subgroup A of M(n) we denote its Lie algebra by L(A). Using the standard notations, L(M(n)) = so(n+1,1), so has R-rank 1. We now assume that G is semisimple and noncompact. So

(4.6.1) L(G) $= \oplus L(H_i)$ where H_i is a simple normal subgroup of G. Clearly

(4.6.2)

$$the\,R - rank\,of\,G = \sum_{i=1}^{r} the\,R - rank\,of\,H_i \leq the\,R - rank\,of\,M(n) = 1.$$

If the R-rank of G were 0, then G would be compact. So the R-rank of G is 1, and for the same reason, precisely one H_i is non-compact. So if $r \geq 2$ then G would contain a compact normal subgroup, and the theorem would be true for G. In other words

Assertion 4: To prove the theorem we may assume that G is simple, and non-compact.

(4.7) A simple real Lie algebra of R-rank 1 is isomorphic to one of the following, cf.[3], p.518, table V:

(4.7.1)

$$i) o(k,1), \quad ii) u(k,1), \quad iii) sp(k,1), \quad iv) f_{4(-20)}, \quad k \geq 2.$$

So L(G) has one of these forms. But L(G) $\leq$ L(M(n)) = so(n+1,1), so it admits a faithful representation on a real vector space V of dimension n+2 and preserves a Lorentz metric on V. From the considerations of weights in finite-dimensional representations of these Lie algebras it is not difficult to see that none of the Lie algebras of types ii), iii) or iv) above admit such a representation. Moreover if so(k,1) acts on V leaving invariant a Lorentz form, then V splits as an so(k,1) - invariant direct sum W $\oplus$ U where the form restricted to W (resp. U) is Lorentz (resp. definite), dim W = k+1 , dim U = n+1-k. By exponentiating we see that $G_\circ$ must be conjugate to $SO_\circ$(k,1) which is embedded in a standard way in M(n). So up to conjugacy, $G_\circ \leq G_\circ(k-1)$, $k \geq 2$. This finishes the proof of Greenberg's theorem. q.e.d.

§5 Extensions of Theorems of Nielsen and Van Vleck

(5.1) Class of a Subgroup Let Γ be a subgroup of M(n),and let G be its closure. If G is compact then we set the *class of* Γ to be -2. If G is noncompact and fixes a unique point in $\mathbf{S}^n$ then we set it to be -1. Otherwise let k be the unique smallest integer ≥ 0 such that G leaves a round k-sphere invariant. In this case we set the class of Γ to be k.

(5.2) Theorem Let Γ be a subgroup of M(n) of class k. Suppose that k is odd ≥ 1, and Γ does not contain elliptic elements accumulating at identity. Then Γ is discrete.

(5.3) Proof Suppose that Γ is not discrete. Let G be its closure and G_o the identity component of G. Suppose first that the class of Γ is actually n. Then by Greenberg's theorem G_o equals $M_o(n)$, i.e. $\Gamma \cap M_o(n)$ is dense in $M_o(n)$. But since in our case n is odd it follows by (2.5) that Γ must contain elliptic elements accumulating at identity. This contradicts our hypothesis. So Γ must be discrete.

Now suppose that the class of Γ is k $<$ n and k is odd ≥ 1. Then again by Greenberg's theorem $G \subset G(k) \approx M(k) \times O(n\text{-}k)$. Let Φ be the projection of Γ in the first factor M(k). Since Γ does not contain elliptic elements accumulating at identity it follows that this projection restricted to Γ has finite kernel and so Φ is a non-discrete subgroup of M(k) of class k containing no elliptic elements accumulating at identity. As above we again arrive at a contradiction. So Γ must be discrete. q.e.d.

(5.4) Remarks The above theorem in case n = 1, and every non-identity element of Γ is hyperbolic is due to Nielsen, cf.[5]. In the above formulation, again for n = 1, it was proved by Siegel, cf.[6]. The above generalization is due to Van Est, cf.[8], and Greenberg, cf.[2]. For further extensions along the same lines see [1], [2].

(5.5) An element g in M(n) is said to be *rotationless* if all of its rotation- angles are zero. An obvious but remarkable property which plays a role in the classical theory of Fuchsian groups is that every non-elliptic element of M(1) $\approx$ PSL(2,R) is rotationless.

(5.6) Theorem Let Γ be a subgroup of M(n) of class $\geq$1, and consisting of rotationless elements only. Then the class of Γ must be 1, and Γ is discrete.

(5.7) Proof Notice that in the linear model of M(n) a rotationless element, if hypebolic, has n eigevalues equal to 1, and if parabolic has all eigenvalues equal to 1. Let G be the Zariski-closure of Γ, i.e. the smallest algebraic subgroup containing Γ. It follows that every element of G has at least n eigenvalues equal to 1. Of course G is a Lie subgroup of M(n). Its class is k which by hypothesis is ≥ 1.So its identity component has a noncompact semisimple part $\approx M_o(k)$. The algebraic property of the eigenvalues forces that we must have k = 1. Moreover Γ clearly does not contain any non-identity elliptic element. So by (5.2) Γ is discrete. q.e.d.

(5.8) Let n = 2. Then M(2) ≈ PSL(2,C) and we can talk of a trace of an element of M(2) – upto sign – as a complex number. We see readily that a non-elliptic element of M(2) is rotationless iff its trace is real. So (5.7) implies that a subgroup of PSL(2,C) all of whose non-identity elements are non-elliptic and have real traces is in fact conjugate to a torsion-free Fuchsian group. In this setup the theorem is due to Van Vleck,cf.[8].

§6 Existence of Hyperbolic Elements

(6.1) Theorem Let Γ be a subgroup of M(n). Suppose Γ does not fix a point in $\bar{\mathbf{D}}^{n+1}$. Then Γ contains a hyperbolic element.

The proof is based on the following fairly standard lemma on linear groups. Its proof is omitted.

(6.2) Lemma Let G be a subgroup of linear transformations of an N-dimensional real vector space V. Suppose that every $g \in G$ has all its eigenvalues of absolute value 1. Then there exists G-invariant subspaces

$$W_\circ = V \supset W_1 \supset W_2 \supset \ldots \supset W_k$$

such that the G-action on $W_{i-1}/W_i, i = 1, \ldots k$ is conjugate to an R-irreducible orthogonal action.

(6.3) Proof of the theorem : Consider the linear model for M(n), cf. chapter 1, §4. So we regard Γ as acting linearly on $V = R^{n+2}$. Suppose Γ does not contain a hyperbolic element. Then Γ satisfies the condition stated in (6.2). Let W be a minimal Γ-invariant subspace of dimension ≥ 1 in V. If W = V then by (6.2) Γ is conjugate to a subgroup of O(n+2), and so Γ has a fixed point in $\mathbf{D}^{n+1}$. So suppose $W \neq V$. If the Lorentz metric $<,>$ on V restricted to W is degenerate then W is tangential to the quadric, and so by minimality dim W = 1. So Γ fixes a point in $\mathbf{S}^n$. So we now suppose that $<,>|_W$ is non-degenerate. For definiteness we take $<,>$ with the signature $- + + + \ldots$. If $<,>|_W$ is negative-definite then dim W = 1 and Γ has a fixed point in $\mathbf{D}^{n+1}$. Otherwise $<,>$ restricted either to W or to its orthogonal complement is a Lorentz metric. Let U be this Γ-invariant proper Lorentz subspace. It defines a $\mathbf{D}^k, 1 \leq k \leq n$ in $\mathbf{D}^{n+1}$. It is clear that g in Γ is hyperbolic iff g restricted $\mathbf{D}^k$ is hyperbolic. So the result follows by induction on n. q.e.d.

(6.4) The above proof also shows the following

Theorem. Let Γ be a subgroup of M(n) consisting only of elliptics and parabolics. Then Γ fixes a point in $\bar{\mathbf{D}}^{n+1}$.

§7 Groups Consisting of Elliptics

(7.1) Let Γ be a subgroup of M(n) consisting only of elliptics. One might conjecture that Γ must have a fixed point in $\mathbf{D}^{\mathbf{n+1}}$, not just in $\bar{\mathbf{D}}^{\mathbf{n+1}}$ as (6.4) asserts, i.e. Γ must be conjugate to a subgroup of O(n+1). The answer is a bit surprising.

(7.2) Theorem Let Γ be a subgroup of M(n) (resp. E(n)) consisting of elliptics only. If

i) Γ contains a solvable subgroup of finite index, *or*

ii) Γ is discrete, *or*

iii) $n \leq 3$,

then Γ has a fixed point in $\mathbf{D}^{\mathbf{n+1}}$ (resp.$\mathbf{E}^{\mathbf{n}}$). If $n \geq 4$ then there exist subgroups of M(n) (resp. E(n)) consisting only of elliptics but having no fixed point in $\mathbf{D}^{\mathbf{n+1}}$ (resp.$\mathbf{E}^{\mathbf{n}}$).

The proof extends over the rest of this section.

(7.3) Notice that in view of (6.4) it suffices to prove the assertion for E(n) acting on $\mathbf{E}^{\mathbf{n}}$. Assume first that Γ is abelian, and g is in Γ, $g \neq e$. Let $\mathbf{E}^{\mathbf{k}}, k \geq 0$, be the fixed point set of g. It is clear that Γ leaves $\mathbf{E}^{\mathbf{k}}$ invariant and its restriction to $\mathbf{E}^{\mathbf{k}}$ is an abelian group consisting of elliptics only. An easy induction on n shows that Γ has a fixed point.

(7.4) Next suppose that Γ has a solvable subgroup of finite index. It is then easy to see that Γ then has a normal solvable subgroup H of finite index. If $H = \{e\}$ then Γ is finite, and a "center of mass" - construction gives a fixed point for Γ. Otherwise H contains a characteristic abelian subgroup $K \neq \{e\}$. By (7.3) K has a non-empty fixed point set, say $\mathbf{E}^{\mathbf{k}}, k > 0$. Then $\mathbf{E}^{\mathbf{k}}$ is left invariant by Γ, and again by induction on n it follows that Γ has a fixed point.

(7.5) Now suppose that Γ is discrete, i.e. Γ is a Bieberbach group. So Γ contains a normal abelian subgroup H of finite index consisting of translations. Since we are assuming that Γ consists of elliptics only we must have $H = \{e\}$, and Γ is finite, so again a "center of mass"-construction gives a fixed point for Γ.

(7.6) Now we suppose that $n \leq 2$. Then E(n) is solvable. So Γ is also solvable,and by (7.4) Γ has a fixed point.

(7.7) Now suppose that n = 3, and Γ non-discrete. Let Γ_o be the identity component of the closure of Γ. Now it is easy to see from Lie-algebra considerations that the only connected, proper, non-solvable subgroups of E(3) are the conjugates of SO(3). So either Γ_o is $E_o(3)$ or a conjugate of SO(3). In the first case, by (1.8) Γ would contain a parabolic element. In the second case Γ_o would have a fixed point which will be also fixed by Γ.

(7.8) Finally suppose that $n \geq 4$. For definiteness let $n = 4$. Now E(4) contains SU(2) –in fact uniquely upto conjugacy– and each non-identity element of SU(2) has both rotation-angles non-zero. Also SU(2) contains non-abelian free subgroups. Let $\Phi =< A, B >$ be such a free subgroup, and set

(7.8.1)

$$\Gamma = \langle a : \vec{x} \mapsto A\vec{x},\, b : \vec{x} \mapsto B\vec{x} + \vec{v}\rangle,\ \vec{v} \neq 0.$$

Let $\rho : E(4) \longrightarrow O(4)$ be the canonical projection, cf (1.3). Then $\rho(a) = A$, and $\rho(b) = B$. So $\rho(\Gamma) = \Phi$. Since Φ is free so is Γ. Observe that each of a and b have a unique fixed point and these fixed points are distinct. So G does not have a common fixed point. On the other hand, for each non-identity element g of Γ, we have $\rho(g) \neq e$. So $\rho(g)$ and hence g have both rotation-angles non-zero. So each g in Γ has a fixed point, so Γ consists of elliptics only.

This finishes the proof of theorem (7.2).

References

1) S.S.Chen and L.Greenberg, Hyperbolic Spaces, Contributions to Analysis,ed. by B.Maskit, Academic Press(1974), 49-87.

2) L.Greenberg, Discrete Subgroups of the Lorentz group, Math.Scand.10(1962), 85-102.

3) S.Helgason, Differential Geometry, Lie Groups, and Symmetric Spaces, Academic Press(1978).

4) R.S.Kulkarni and F.Raymond, 3-Dimensional Lorentz Space-forms and Seifert Fiber Spaces, Jour.of Diff.Geom.21 (1985), 231-268.

5) J.Nielsen, Über Gruppen linearer Transformationen, Mitt.Math.Gesellsch Hamburg 8(1940), 82-104.

6) C.L.Siegel, Bemerkung zu einem Satz von Jacob Nielsen, Mat.Tidsskr.B(1950), 66-70.

7) W.T.Van Est, A Generalization of a Theorem of J.Nielsen concerning Hyperbolic Groups, Ph.D.Dissertation, Univ.of Utrecht, 1950.

8) E.B.Van Vleck, On the Combination of Non-loxodromic Substitutions, Trans. Amer.Math.Soc.20(1919), 299-317.

Conformal Geometry from the Riemannian Viewpoint

Jacques Lafontaine

Contents

A. Introduction

When a conformal structure on a manifold is defined by a Riemannian metric g, how to detect conformal flatness on g ? The answer, due to Weyl and Schouten, is given in § C, and some applications are derived in § D. It turns out that the three dimensional case, i.e. the case where the curvature tensor is determined by the Ricci tensor, needs a special treatment. An example of that situation is given in § E. We also give some global properties of compact conformally flat manifolds: the nullity of their Pontryagin numbers, (Chern-Simons), a vanishing theorem for middle-dimensional cohomology when the scalar curvature is positive (Bourguignon) and a structure theorem when the scalar curvature is zero.

B. Curvature via representation theory.

Let (E,q) be a real vector space with dimension $n > 1$, equipped with a non degenerate quadratic form q . Of course, we are thinking of $(E,q) = (T_m M, g_m)$ for a (pseudo)-Riemannian manifold (M,g).

There is a natural action of $GL(E)$ on each tensor space $\otimes^k E \otimes^\ell E^*$. Indeed, for $x_i \in E$ $(1 \leq i \leq k)$, $y_j \in E^*$ $(1 \leq j \leq \ell)$ and $\gamma \in GL(E)$, set

$$\gamma(x_1 \otimes \ldots \otimes x_k \otimes y_1 \ldots \otimes y_\ell) = \gamma \cdot x_1 \otimes \ldots \quad {}^t\gamma^{-1} y_\ell .$$

Since this action preserves the symmetries that a given tensor may possess, it is not irreducible. For instance, $\otimes^2 E$ clearly admits the $GL(E)$-decomposition

$$\otimes^2 E = S^2 E \oplus \Lambda^2 E$$

(the symmetric product being denoted by $S^2 E$. This decomposition is irreducible, cf. [W2].)

The decomposition of $\otimes^k E$ is more involved. It uses the natural action of the symmetric group S_k on $\otimes^k E$, of the group algebra $\mathbb{R}[S_k]$, and certain idempotents of $\mathbb{R}[S_k]$, the so-called Young symmetrizers, cf. [W2], ch. IV or [N-S], II.3.

Now, q gives an identification of E and E^*. If $\gamma \in O(q)$, then $\gamma = {}^t\gamma^{-1}$, which proves that the $O(q)$-modules E and E^* are isomorphic. From now on, only tensor powers of E^* will be considered. The simplest case to consider is $\otimes^2 E^*$. We have

$$(1) \qquad \otimes^2 E^* = \mathbb{R} \cdot q \oplus S_0^2 E^* \oplus \Lambda^2 E^* \quad .$$

Here, $S_0^2 E^*$ is the space of symmetric two-tensors whose trace is zero; clearly, the action of $O(q)$ on the one-dimensional factor $\mathbb{R} \cdot q$ is trivial.

The reader may look at [Ha] § 11 for a remarkable application to Riemannian geometry of the decomposition of $E^* \otimes S^2 E^*$.

Now, our main concern is the curvature tensor $R \in \otimes^4 E^*$ $(E = T_m M)$. The identities

$$R(x,y,z,t) = -R(y,x,z,t) = -R(x,y,t,z)$$

and

$$R(x,y,z,t) = R(z,t,x,y)$$

just mean that $R \in S^2(\Lambda^2 E^*)$. Now, it can be checked that the only idempotent of $\mathbb{R}[S_4]$ whose restriction to $S^2(\Lambda^2 E^*)$ is not trivial is the <u>Bianchi map</u>, which is given by

$$b(R)(x,y,z,t) = \frac{1}{3}[R(x,y,z,t) + R(y,z,x,t) + R(z,x,y,t)].$$

This proves by the way that $S^2(\Lambda^2 E^*)$ admits, with respect to the action of $GL(E)$, the irreducible decomposition

$$S^2(\Lambda^2 E^*) = \operatorname{Ker} b \oplus \operatorname{Im} b.$$

Now, since $b(\alpha \circ \beta) = \frac{1}{6}\, \alpha \wedge \beta$, $\operatorname{Im} b$ is $GL(E)$-isomorphic to $\Lambda^4 E^*$. In particular, $b \equiv 0$ if $\dim E = 2$ or $= 3$.

2. <u>Definition.</u> The vector space (and $O(q)$-module) $\operatorname{Ker} b$ is the space of <u>curvature tensors.</u>

It will be denoted by $\mathcal{C}E$. We already know that

$$\begin{aligned} \dim\ \mathcal{C}E &= \dim S^2(\Lambda^2 E^*) - \dim \Lambda^4 E^* \\ &= n^2(n^2 - 1)/12 \quad . \end{aligned}$$

There is an $O(q)$-equivariant map of $\mathcal{C}E$ into $S^2 E^*$, the Ricci-contradiction, defined by setting

$$(3) \qquad c(R)(x,y) = \sum_{i=1}^{n} R(x,e_i,y,e_i) \quad ,$$

where $(e_i)_{1 \leq i \leq n}$ is an orthonormal basis of (E,q).

There is also a natural way of generating elements of $\mathcal{C}E$ with symmetric two-tensors.

4. <u>Definition.</u> Given $h,k \in S^2 E^*$, the <u>Kulkarni-Nomizu</u> product $h \cdot k \in \mathcal{C}E$ is given by

$$(h \cdot k)(x,y,z,t) = h(x,z)k(y,t)$$

$$+ h(y,t)k(x,z) - h(x,t)k(y,z) - h(y,z)k(x,t).$$

Examples i) If a,b,c,d are one-forms,

$$(a \circ b) \cdot (c \circ d) = (a \wedge c) \otimes (b \wedge d)$$

ii) The tensor $g \cdot g$ is just the double of the curvature tensor of the standard sphere.

iii) If $h = \sum \lambda_i (e_i \circ e_i)$ is diagonalized with respect to an orthonormal basis, then

$$g \cdot h = 2 \sum_{i<j} (\lambda_i + \lambda_j)(e_i \wedge e_j) \otimes (e_i \wedge e_j) \quad .$$

In the following, we shall need $0(q)$-invariant scalar products on S^2E^* and $S^2(\Lambda^2E^*)$. To get these, embed in the natural way, using q, S^2E^* and $S^2(\Lambda^2E^*)$ into End E and End Λ^2E respectively, and define the scalar product of two endomorphisms to be trace of their product.

5. Lemma. i) the map $h \mapsto h \cdot q$ of S^2E^* into $\mathcal{C}E$ is $0(q)$-equivariant, and its transposed map is the Ricci contradiction c.

ii) if $n > 2$, it is injective.

Proof. i) is straightforward. For ii), remark that
$c(h \cdot q) = (n-2)h + (\mathrm{tr}_q h)q$

6. Theorem. i) if $n = 3$, the $0(q)$-module $S^2(\Lambda^2E^*)$ (equal to $\mathcal{C}E$ in that case) is $0(q)$-isomorphic to S^2E^*.

ii) if $n \geqq 4$, the $0(q)$-module $S^2(\Lambda^2E^*)$ admits the following irreducible decomposition

$$S^2(\Lambda^2E) \cong \mathbb{R} \oplus S^2_0E^* \oplus \mathcal{W}E \oplus \Lambda^4E^* \; .$$

Here $\mathbb{R}$ and S^2_0E are realized in $S^2(\Lambda^2E^*)$ by $\mathbb{R}\, q \cdot q$ and

$q \cdot S_0^2E^*$ <u>respectively</u>; $\mathcal{W}E = \operatorname{Ker} c \cap \operatorname{Ker} b$.

<u>Proof</u> (summarized). i) If $n = 3$, the injection $h \longmapsto h \cdot q$ is an isomorphism for dimension reasons.

ii) Using the lemma B.5 and (1), we see that $S^2(\Lambda^2 E)$ admits the $O(q)$-invariant subspaces $\mathbb{R}\, q \cdot q$, $q \cdot S_0^2 E$ and $\operatorname{Im} b$, which are $O(q)$-isomorphic to $\mathbb{R}$, S_0^2E and $\Lambda^4 E$ respectively. Clearly, the orthogonal space to the direct sum of these three spaces is $\operatorname{Ker} c \cap \operatorname{Ker} b$, using the lemma again.

Now, we have to check irreducibility. On one hand (see [SB], n°IX for further details) the vector space of $O(q)$-invariant quadratic forms on $\mathcal{C}E$ is three-dimensional, and generated by $|R|^2$, $|c(R)|^2$ (the quadratic forms associated with the scalar products on $S^2(\Lambda^2E^*)$ and S^2E^* we defined earlier) and $(\operatorname{tr}_q c(R))^2$. So $\mathcal{C}E$ admits at most three $O(q)$-irreducible components. Since we have found three components, they are indeed irreducible. On the other hand Λ^4E^* (and also any Λ^kE^*) is $O(q)$-irreducible, since the space of $O(q)$-invariant quadratic forms on Λ^kE^* is one-dimensional. □

7. <u>Definition.</u> $\mathcal{W}E$ is the space of <u>Weyl curvature tensors</u> associated with (E,q).

This definition still makes sense when $n = 3$, but then $\mathcal{W} = 0$. We shall use the following corollary of theorem B.6.

8. <u>Corollary.</u> <u>Given</u> $R \in \mathcal{C}E$ $(\dim E \geq 3)$ <u>there exist unique</u> $h \in S^2E^*$ <u>and</u> $W \in \mathcal{W}E$ <u>such that</u>

$$R = q \cdot h + W.$$

When R is the curvature tensor of a Riemannian metric, taking the Ricci-contradiction of both sides we get $\operatorname{Ric} = (n-2)h + (\operatorname{tr} h)g$ and therefore

$$h = \frac{1}{n-2}\left[\operatorname{Ric} - \frac{s}{2(n-1)}\, g\right]$$

As for W, it can be viewed as the "remainder" after a division by q (or g).

9. Definition. We call h and W the Schouten tensor and the Weyl curvature of the metric g.

C. Detecting conformal flatness: the Weyl-Schouten theorem.

The following convention is useful: we denote vector fields by capital letters X, Y, Z etc. and vectors by small letters. For instance, to put some emphasis on the properties of connections, at a given point we shall write $D_x Y$ instead of $D_X Y$. Recall that the difference of two connections D and D' is a tensor of type $(1,2)$, which is symmetric if both connections are torsion free. So we write at each point

$$D'_y Z = D_y Z + C(y,z)$$

1. Proposition. *The curvatures R and R' of D and D' satisfy the relation*

$$R'(x,y)z = R(x,y)z - (D_x C)(y,z) + (D_y C)(x,z)$$

$$- C(x,C(y,z)) + C(y,C(x,z)).$$

Proof. Straightforward computation. Recall that $(D_x C)(y,z)$ is the covariant derivative of C : it is a tensor of type $(1,3)$.

2. Examples. i) If D and D' are projectively equivalent (i.e. have the same geodesic paths), it is well known (cf. [Sp], II.6.35 for instance) that $C(y,z) = \alpha(y)z + \alpha(z)y$, where α is a 1-form. Then

$$R'(x,y)z = R(x,y)z - d\alpha(x,y)z$$

$$-(D_x\alpha)(z)y + (D_y\alpha)(z)x - \alpha(y)\alpha(z)x + \alpha(x)\alpha(z)y$$

ii) If D and D' are the Levi-Civita connections of two point-wise conformal metrics g and $g' = e^{2f}g$, we have seen in ch. I § 2 that

$$C(y,z) = df(y)z + df(z)y - g(y,z)\nabla f .$$

It follows that

$$\begin{aligned} R'(x,y)z &= R(x,y)z - (Ddf)(x,z)y \\ &+ (Ddf)(y,z)x + g(y,z)D_x\nabla f \\ &- g(x,z)D_y\nabla f - df(z)\ [df(y)x - df(x)y] \\ &+ |df|^2[g(y,z)x - g(z,x)y] \\ &+ [g(x,z)\alpha(y) - g(y,z)df(x)]\nabla f \end{aligned}$$

If we compare the (0,4) tensors instead and use the formalism of the preceeding paragraph, this formula gets nicer. Indeed

3. <u>Proposition.</u> <u>Let</u> R <u>and</u> R' <u>the</u> (0,4) <u>curvature tensors of the metrices</u> g <u>and</u> $g' = e^{2f}g$ <u>respectively. Then</u>

$$e^{-2f}R' = R - Ddf \cdot g + (df \circ df) \cdot g - \frac{1}{2}|df|^2\, g \cdot g$$

<u>Proof.</u> Use the very definition of our product on two-forms.

□

In particular, the Weyl curvatures and the Schouten tensors satisfy, using corollary B.8,

$$W' = e^{2f}W$$

$$h' = h - Ddf + df \circ df - \frac{1}{2}|df|^2_g .$$

4. <u>Remarks.</u> i) Coming back to (1,3) tensors, we see that $W' = W$ in that case. That's why W is often called the conformal curvature.

(ii) We can deduce from these formulae the relations between the Ricci and the scalar curvature. It turns out that it is then more convenient to write the conformal factor differently. Indeed, setting $g' = \varphi^{-2} g$, we get

$$\mathrm{Ric}' - \frac{s'}{n} g' = \mathrm{Ric} - \frac{s}{n} g + \frac{(n-2)}{\varphi}(Dd\varphi + \frac{\Delta\varphi}{n} g)$$

Setting $g' = \Psi^{\frac{4}{n-2}} g$, we get

$$\Psi^{\frac{n+2}{n-2}} \ \mathrm{Scal}' = \mathrm{Scal}\ \Psi + 4\ \frac{(n-1)}{n-2}\ \Delta\Psi\ .$$

iii) If $n = 2$, we get $\mathrm{Ric}' = \mathrm{Ric} + (\Delta f)g$, therefore the Gauss curvatures K and K' satisfy

$$e^{2f}K' = K + \Delta f\ .$$

We can now answer the question of detecting conformal flatness. For the two-dimensional case, we have the well-known theorem of existence of isothermal coordinates, proved by Gauss in the C^{ω} case.

5. Theorem. Any two-dimensional C^{∞} Riemannian manifold (M,g) is conformally flat.

Proof. We must find, in the neighbourhood of each point, a function f such that the Gaussian curvature of $e^{2f}g$ be zero. It amounts to solve the elliptic equation $\Delta f + K = 0$, which always has local solutions. □

When $\dim M \geq 3$ an obvious necessary condition for conformal flatness is the nullity of the Weyl tensor, since it is conformally invariant up to a mutliplicative factor. Of course, this condition is always satisfied if $\dim M = 3$. Now, we must find locally a function f such that the Schouten tensor of the metric $e^{2f}g$ be zero, i.e. solve the over-determined equation

$$h - Ddf + df \circ df - \frac{1}{2}\ |df|^2 g = 0$$

6. Technical lemma. *Let* (M,g) *a Riemannian manifold* ($\dim M \geq 3$) *with zero Weyl curvature. Then* g *is conformally flat if and only if*

$$D_x h(y,z) - D_y h(x,z) = 0$$

(i.e. iff the covariant derivative of the Schouten tensor is completely symmetric.)

Proof. If g is conformally flat, each point of M admits a neighbourhood U such that the equation

$$Ddf - df \circ df + \frac{1}{2}|df|^2 g = h$$

admits a solution f.

It amounts to the same to say that there exists locally a one form α such that

$$D\alpha - \alpha \circ \alpha + \frac{1}{2}|\alpha|^2 g = h \tag{7}$$

Indeed, for such an α the covariant derivative $D\alpha$ is symmetric. This means that α is closed, and since the problem is local, we can apply Poincaré lemma.

Now, define $d^D : C^\infty(S^2M) \longrightarrow C^\infty(\Lambda^2 M \otimes TM)$

$$d^D s(x,y,z) = D_x\, s(y,z) - D_y s(x,z). \tag{8}$$

Applying d^D to both members of (7) and taking in account that α must be closed, we get

$$d^D h(x,y,z) = R(x,y,z,\alpha) - \alpha(y) D_x\alpha(z)$$

$$+ \alpha(x) D_y\alpha(z) + g(D_x\alpha,\alpha) g(y,z)$$

$$- g(D_y\alpha,\alpha) g(x,y).$$

Taking (7) into account, we get

$$d^D h(x,y,z) = R(x,y,z,\alpha)$$
$$-\alpha(y)[\alpha(n)\alpha(z) - \frac{1}{2}|\alpha|^2 g(x,z) - h(x,z)]$$
$$+\alpha(x)[a(y)\alpha(z) - \frac{1}{2}|\alpha|^2 g(y,z) - h(y,z)]$$
$$+g(y,z)[\frac{1}{2}|\alpha|^2\alpha(x) + h(x,\alpha)]$$
$$-g(x,z)[\frac{1}{2}|\alpha|^2\alpha(y) + h(y,\alpha)]$$

i.e. $$d^D h(x,y,z) = R(x,y,z,\alpha)$$
$$+\alpha(y)h(x,z) - \alpha(x)h(y,z) + h(x,\alpha)g(y,z)$$
$$-h(y,\alpha)g(x,z).$$

Now, $R = g \cdot h + W$. So we get

$$d^D h(x,y,z) = W(x,y,z,\alpha) = 0$$

since $W = 0$.

Conversely, suppose that $W = 0$ and $d^D h = 0$. Then the equation $D\alpha - \alpha \circ \alpha + \frac{1}{2}|\alpha|^2 g = h$ can be viewed (using local coordinates for instance) as an overdetermined system of the type

$$\partial_i \alpha_j = f_{ij}(\alpha_1, \ldots, \alpha_n) + h_{ij}$$

and the conditions $d^D h = 0$ and $W = 0$ just say, mimicking the computation above, that this system is completely integrable. (See [Sp] , I.6 for this less classical form of the Frobenius theorem).

□

9. Theorem. (Weyl-Schouten) i) *A 3-dimensional Riemannian manifold (M,g) is conformally flat if and only if its Schouten tensor satisfies*

$$D_x h(y,z) - D_y h(x,z) = 0$$

ii) *If* $\dim M \geq 4$, *then* (M,g) *is conformally flat if and only if* $W = 0$.

<u>Proof</u>. i) is contained in the technical lemma. As for ii), what remains to be proved is that, when $\dim M \geq 4$, the condition $d^D h = 0$ is a consequence of $W = 0$. This comes from the second Bianchi identity. Indeed, since $W = 0$, $R = g \cdot h$ and $D_x R(y,z,t,u) = g(y,t) D_x h(z,u) + g(z,u) D_x h(g,t)$ $- g(z,t) D_x h(y,u) - g(y,u) D_x h(z,t)$.
Therefore

$$g(x,u)\, d^D h(y,z,t) + g(y,u)\, d^D h(z,x,t)$$

$$+ g(z,u)\, d^D h(x,y,t) - g(x,t)\, d^D h(y,z,u)$$

$$- g(y,t)\, d^D g(z,x,u) - g(z,t)\, d^D h(x,y,u) = 0$$

Taking the trace with respect to x,u we get

$$(n-3)\, d^D h(y,z,t) - g(y,t)\,[\operatorname{div} h(z) + d\operatorname{tr} h(z)]$$

$$+ g(z,t)\,[\operatorname{div} h(y) + d\operatorname{tr} h(y)] = 0$$

(We have denoted by δ the divergence of a symmetric two-tensor: $\operatorname{div} h(x) = - \Sigma D_{e_i} h(x,e_i)$ for an orthonormal basis $(e_i)_{1 \leq i \leq n}$). Now the second Bianchi identity says that $\operatorname{div} \operatorname{Ric} = - \frac{1}{2} d \operatorname{Scal}$, which just means that $\operatorname{div} h + d\operatorname{tr} h = 0$. Of course, this relation can be obtained by taking a further trace with respect to z,t.

□

<u>Remarks.</u> i) The case of projectively flat connections can be treated by a quite analogous two-step process.

ii) It would be tempting to state and prove some general Frobenius theorem involving the equation $D\alpha = F(\alpha)$. Unfortunately, if we want F to be "natural" in a reasonable sense, such a theorem will only cover the equations obtained in the projective and in the conformal case.

D. Some consequences: Riemannian products, conformally flat hypersurfaces of $\mathbb{R}^n$, $n \geq 5$.

As a consequence of theorem C.9 above, we can give a list of examples of conformally flat manifolds. In dimension 3, every Riemannian manifold with parallel Ricci tensor is conformally flat (this condition is not necessary of course, cf. E section). Not surprisingly, we get three dimensional space-forms, and also products of two dimensional space forms with the real line.

In dimension greater than or equal to four, we have to check the "divisibility" of the Riemann curvature tensor by the metric g. The following example is very easy to get

1. Proposition. i) If (U_+,g_+) and (U_-,g_-) are two space-forms where curvatures are $+1$ and -1 respectively, the Riemannian product $(U_+ \times U_-, g_+ \times g_-)$ is conformally flat.

ii) If (U,g) is a space form, the product $(U \times I,\ g + dt^2)$ is conformally flat, and so is the warped product $(U \times I,\ f^2(t)g + dt^2)$ for any non vanishing function f on I.

Proof. i) The curvature tensors of g_+ and g_{-1} are $\frac{1}{2}g_+ \cdot g_+$ and $-\frac{1}{2}g_- \cdot g_-$ respectively. The curvature of the product (abusively written $g = g_+ + g_-$) is just

$$R = R_+ + R_- = \frac{1}{2}(g_+ \cdot g_+ - g_- \cdot g_-)$$

$$= \frac{1}{2}(g_+ + g_-) \cdot (g_+ - g_-) = \frac{1}{2}\, g \cdot (g_+ - g_-)$$

since the product is commutative and distributive.

ii) The first part goes is the same way: the curvature of the product is the same as the curvature of the first factor, i.e.

$$K\, g \cdot g = K(g + dt^2) \cdot (g - dt^2),$$

since $dt^2 \cdot dt^2 = 0$.

As for the metric $f^2(t)g + dt^2$, we can write

$$f^2(t)g + dt^2 = f^2(t)\left[g + \left(\frac{dt}{f(t)}\right)^2\right]$$

$= \varphi^2(u)[g + du^2]$ after change of variable, and use the conformal flatness of $g + du^2$.

□

Remark. These results can be checked directly, without using Weyl's theorem. Indeed, consider $\mathbb{R}^{p+q} - \mathbb{R}^{q-1}$, which is diffeomorphic to $S^p \times \mathbb{R}^q$, equipped with the metric

$$\left(\sum_{i=1}^{p+q}\right) dx_i^2 \;/\; r^2 = \quad \text{where} \quad r^2 = \sum_{i=1}^{p+1} x_i^2$$

Denoting by $d\sigma_p$ the standard metric of $S^p \subset \mathbb{R}^{p+1}$, this metric can be written

$$(dr^2 + r^2 d\sigma_p + \sum_{p+2}^{p+q} dx^2)/r^2 =$$

$$d\sigma_p + (dr^2 + \sum_{p+2}^{p+q} dx_i^2)/r^2$$

The second term is just the Poincaré metric of the half space $\mathbb{R}^+ \times \mathbb{R}^{q-1}$. Compare with chapter I, § 7.

The following property provides a further opportunity of playing with divisibility by g, but has also global consequences we will see later.

2. Proposition. A Riemannian product $(M_1, g_1) \times (M_2, g_2)$ is conformally flat if and only if either

i) (M_i, g_i) is one dimensional, and (M_j, g_j) $(j \neq i)$ a space-form, or

ii) (M_1, g_1) and (M_2, g_2) are space-forms of dimension at least two, with non-zero opposite curvatures.

Proof. The "if" part has been checked already. Now, the curvature of the product metric is

$$g_1 \cdot h_1 + g_2 \cdot h_2 + W_1 + W_2 .$$

Clearly, W_1 and W_2 belong to the Weyl component for the product metric, and must be zero if the metric is conformally flat. So we must decide when $g_1 h_1 + g_2 h_2$ is divisible by $g = g_1 + g_2$. In that case, since

$$c(g_i \cdot h_i) = (n_i - 2)h_i + (\mathrm{tr}\, h_j)g_i$$

$$(i = 1,2,\ n_i = \dim M_i)\ ,$$

$$(n - 2)(h_1 \cdot g_1 + h_2 \cdot g_2) = [(n_1 - 2)h_1 + (n_2 - 2)h_2$$

$$+ c_1 g_1 + c_2 g_2] \cdot (g_1 + g_2),$$

where c_i are scalars. Projecting on each factor, we get

$$(n-2)h_i \cdot g_i = ((n_i - 2)h_i + c_i g_i) \cdot g_i\ ,$$

therefore

$$(n-2)h_i = (n_i - 2)h_i + c_i g_i$$

if $n_i > 2$. Since $n \neq n_i$, this proves that each factor is Einstein, and consequently a space-form, since we already know it is conformally flat.

Then $2R = K_1 g_1 \cdot g_1 + K_2 g_2 \cdot g_2$

$$= (g_1 + g_2)\cdot(K_1 g_1 + K_2 g_2) - (K_1 + K_2) g_1 \cdot g_2$$

Now, it is easy to check that $g_1 \cdot g_2$ is not divisible by $g_1 + g_2$, unless one g_i is equal to dt^2.

Analogous proofs work for $n = 3$ or $n = 4$ and $n_1 = n_2 = 2$, by a more direct argument.

□

Using our algebraic formalism, it is easy to recover the following property, first proved by Elie Cartan.

3. Theorem. *If $n \geq 4$, a hypersurface $M^n \subset \mathbb{R}^{n+1}$ is conformally flat if and only if the second fundamental form has at each point an eigenvalue of multiplicity at least $n-1$.*

Proof. Let s be the second fundamental form. The Gauss equation can be written $R = \frac{1}{2} s \cdot s$. Sufficiency is then straightforward. Indeed, at each point we have $s = \alpha \circ \alpha + bg$ (the one-form α and the scalar b can of course be zero), therefore, since $(\alpha \circ \alpha) \cdot (\alpha \circ \alpha) = 0$, $s \cdot s = b^2 g \cdot g + 2bg \cdot \alpha \circ \alpha$ is divisible by g.

Conversely, if $s.s$ is divisible by g, since $c(s.s) = 2(\operatorname{tr} s)s - 2s^2$, one has

$$s \cdot s = \frac{1}{(n-2)} [2(\operatorname{tr} s)s - 2s^2 - ag] \cdot g$$

Let $s = \sum_{i=1}^{n} \lambda_i (e_i \circ e_i)$ the spectral decomposition of s. Then, using the very definition of the product. (cf. D.4),

$$s \cdot s = 2 \sum \lambda_i \lambda_j \; (e_i \wedge e_j) \otimes (e_i \wedge e_j) \; .$$

On the other hand, if $h = \sum M_i \, e_i \circ e_i$, then $g \cdot h = 2 \sum_{i<j} (M_i + M_j)(e_i \wedge e_j) \otimes (e_i \wedge e_j)$.
Therefore

$$(n-2)\lambda_i \lambda_j = a'(\lambda_i + \lambda_j) - 2(\lambda_i^2 + \lambda_j^2) - 2a$$

(a and a' are scalar). If $n \geq 4$, a combinatorial argument shows that the λ_i take at most 2 different values. Once this is known, check directly that one of them has multiplicity at least $n-1$. Indeed, if p and q are two orthogonal projectors such that $p + q = \mathrm{Id} = g$ it amounts to check that $p \cdot q$ is never divisible by g, unless one of them has rank one (compare with D.2). □

To go further, one uses the following folk-property.

4. Proposition. *Let h be a tensor on a Riemannian manifold (M,g), satisfying the "Codazzi equation"*

$$d^D h(x,y,z) \equiv D_x h(y,z) - D_y h(x,z) = 0.$$

Let U *be an open set where* h *has an eigenvalue* λ *of constant multiplicity greater than* 1*. Then the corresponding eigen-space distribution is integrable, and* λ *is constant on each leaf.*

Proof. We use the endomorphism field associated with h, and we denote it in the same way. If X and Y are two vector-fields such that $h(X) = \lambda X$ and $h(Y) = \lambda Y$, then

$$(D_Y h)(X) = Y \cdot (h(X)) - h(D_Y X)$$
$$= (Y \cdot \lambda)X + \lambda D_Y X - h(D_Y X)$$

Since $(D_Y h(X) = (D_X h)(Y)$,

$$h([X,Y]) = \lambda[X,Y] + (X \cdot \lambda)Y - (Y \cdot \lambda)X .$$

Using the fact that h is symmetric, we get $h([X,Y]) = \lambda[X,Y]$ and $X \cdot \lambda = Y \cdot \lambda = 0$.

□

Using this result, one can prove (cf. D - D - M) that if M is everywhere non-umbilic, it is foliated by constant curvature manifolds of dimension $n - 1$. Together with D.3, this result is the point of departure of the deeper classification of compact conformally flat hypersurfaces of $\mathbb{R}^n$ $(n \geq 5)$, cf. [C - D - M] and chapter 4.

E. Local theory of conformally flat hypersurfaces of $\mathbb{R}^4$

The case of hypersurfaces of $\mathbb{R}^4$ is more involved, since we are obliged to use the third order condition.
We first prove the following, which is easy, but without any written proof (although implicit in Elie Cartan).

1. *Proposition.* *A hypersurface of* $\mathbb{R}^4$ *whose second fundamental form has at most two eigenvalues is conformally flat.*

Proof. By a continuity argument, it is enough to proof the conformal flatness for open umbilic sets of M, and for open sets where the second fundamental form s has two eigenvalues. The first case is trivial (since M has constant curvature there).

In the second case, write

$$s = \alpha \circ \alpha + fg.$$

Then the Codazzi equation and proposition D.4 imply that $d\alpha \wedge \alpha = 0$ (integrability of the dimension two eigen-distribution) and $df \wedge \alpha = 0$ (constancy of the eigenvalue along the leaves).
Now, $s \cdot s = f^2 g \cdot g + 2f(\alpha \circ \alpha) \cdot g$, so that $\mathrm{Ric}_M = 2f^2 g + |\alpha|^2 fg + f(\alpha \circ \alpha)$, and the Schouten tensor $h = \mathrm{Ric} - \frac{\mathrm{Scal}}{4} g$ is equal to $f(\alpha \circ \alpha) + \frac{1}{2} f^2 g$. Using the equations $d\alpha \wedge \alpha = 0$ and $df \wedge \alpha = 0$, we have

$$d^D h(x,y,z) = f[D_x\alpha(z)\alpha(y) - D_y\alpha(z)\alpha(x)]$$

$$+ f\, df(x)g(y,z) - fdf(g)g(x,z) \quad \text{and}$$

$$0 = d^D s(x,y,z) = D_x\alpha(z)\alpha(y) - D_y\alpha(z)\alpha(x)$$

$$+ df(x)\, g(y,z) - df(y)g(x,z), \quad \text{so that}$$

$$d^D h = fd^D s = 0 .$$

□

To see that this condition is not sufficient, look at the following local example.

Example. Let $M \subset S^3$ (resp. $M \subset \mathbb{R}^3$) be a surface with constant Gaussian curvature, and take the cone $\mathbb{R}^+ M$ in $\mathbb{R}^4$ (resp. the cylinder $\mathbb{R} \times M$). This hypersurface is conformally flat: since its metric is just $dr^2 + r^2 g_M$ (resp. $dt^2 + g_M$) we can use proposition D.1. Now, there exist (see for instance [Sp, III.3]) plenty of local examples of non-umbilical such M - note that this phenemenon is special to dimension 2 - , and in that case the second fundamental form of $\mathbb{R}^+ M$ or

$\mathbb{R} \times M$ has three distinct eigenvalues.

Call generic a hypersurface $M^n \subset \mathbb{R}^{n+1}$ whose second fundamental form has everywhere distinct eigenvalues. The local theory of generic conformally flat hypersurfaces of $\mathbb{R}^4$ is due to Elie Cartan, see [C] p. 89 - 95. We present it here in a somewhat modified way.

For a generic hypersurface, the metric g and the second fundamental form s can be simultaneously diagonalized, so that

$$g = \alpha^2 + \beta^2 + \gamma^2$$

$$s = \lambda\alpha^2 + \mu\beta^2 + \nu\gamma^2 \quad .$$

Furthermore, the functions λ, μ, ν and the one-forms α, β, γ are smooth. We shall use the moving coframe α, β, γ. The Gauss equation says that

$$R = \frac{1}{2}\, s.s = \lambda\mu\ \alpha\wedge\beta\otimes\alpha\wedge\beta + \mu\nu\ \beta\wedge\gamma\otimes\beta\wedge\gamma$$
$$+\ \nu\lambda\ \alpha\wedge\gamma\otimes\alpha\wedge\beta \ ,$$

therefore the Schouten tensor h is given by

$$2h = \lambda_1\alpha^2 + \mu_1\beta^2 + \nu_1\gamma^2 \ ,$$

where $\lambda_1 = \lambda\mu + \lambda\nu - \mu\nu$, $\nu_1 = \nu\lambda + \mu\nu - \lambda\nu$, $\mu_1 = \mu\lambda + \nu\mu - \lambda\mu$.

To work with the conformal flatness criterion namely $d^D h = 0$, we must take into account the Codazzi-equation, which says that $d^D s = 0$.

Let $\begin{pmatrix} 0 & \gamma' & -\beta' \\ -\gamma' & 0 & \alpha' \\ \beta' & -\alpha' & 0 \end{pmatrix}$ be the connection matrix of the coframe α, β, γ. If $\alpha^{\#}, \beta^{\#}, \gamma^{\#}$ are the vector fields associated with α, β, γ be the metric, we have (cf. [Sp],T.2)

$$\alpha'(x) = D_x\gamma(\beta^{\#}), \beta'(x) = D_x\alpha(\gamma^{\#}), \quad \gamma'(x) = D_x\beta(\alpha^{\#}) .$$

Using orthogonality, we also have $D_x\beta(\alpha^{\#}) + D_x\alpha(\beta^{\#}) = 0$, etc...

A straightforward computation gives

$$d^D s(x,y,z) = (d\lambda \wedge \alpha)(x,y)\alpha(z)$$

$$+ (d\mu \wedge \beta)(x,y)\,\beta(z) + (d\nu \wedge \gamma)(x,y)\gamma(z)$$

$$+ \lambda d\alpha(x,y)\alpha(z) + \mu d\beta(x,y)\beta(z) + \nu d\gamma(x,y)\gamma(z)$$

$$+ \lambda(D_x\alpha(z)\alpha(y) - D_y\alpha(z)\alpha(x))$$

$$+ \mu(D_x\beta(z)\beta(y) - D_y\beta(z)\beta(x)$$

$$+ \nu(D_x\gamma(z)\gamma(y) - D_y\gamma(z)\gamma(x))$$

and the same formula for $d^D f$. Taking $z = \alpha^{\#}, \beta^{\#}, \gamma^{\#}$, we see that the Codazzi equations are given by the three following equations, involving two-forms:

$$d\lambda \wedge \alpha + \lambda d\alpha + \mu\gamma' \wedge \beta - \nu\beta' \wedge \gamma = 0$$

$$d\mu \wedge \beta + \mu d\beta + \nu\alpha' \wedge \gamma - \lambda\gamma' \wedge \alpha = 0$$

$$d\nu \wedge \gamma + \nu d\gamma + \lambda\beta' \wedge \alpha - \mu\alpha' \wedge \beta = 0$$

The conformal flatness condition is given by analogous equations, where λ, μ, ν are replaced by λ_1, μ_1, ν_1 .

2. Proposition. *If a generic $M^3 \subset \mathbb{R}^4$ is conformally flat, the "co-principal" distributions $\alpha = 0$, $\beta = 0$, $\gamma = 0$ are integrable.*

Proof. Since we simultaneously have

$$d\lambda \wedge \alpha + \lambda d\alpha + \mu\gamma' \wedge \beta - \nu\beta' \wedge \gamma = 0$$

$$d\lambda_1 \wedge \alpha + \lambda_1 d\alpha + \mu'\gamma \wedge \beta - \nu_1\beta' \wedge \gamma = 0$$

$$d\alpha + \gamma' \wedge \beta - \beta' \wedge \gamma = 0$$

the determinant

$$\begin{vmatrix} d\lambda \wedge \alpha + \lambda\, d\alpha & \mu & \nu \\ d\lambda_1 \wedge \alpha + \lambda_1 d\alpha & \mu_1 & \nu_1 \\ d\alpha & 1 & 1 \end{vmatrix} \text{ is zero.}$$

Since $\begin{vmatrix} \lambda & \mu & \nu \\ \lambda_1 & \mu_1 & \nu_1 \\ 1 & 1 & 1 \end{vmatrix} = (\lambda - \mu)(\mu - \nu)(\nu - \lambda)$

is not zero, the two form $d\alpha$ is divisible by α.

Of course, this condition is not sufficient, since it is satisfied by any $N \times \mathbb{R} \subset \mathbb{R}^4$, where N is any submanifold of $\mathbb{R}^3$. Setting $d\lambda = \lambda_\alpha \alpha + \lambda_\beta \beta + \lambda_\gamma \gamma$, etc...., we get the following criterion.

3. <u>Proposition.</u> <u>A generic $M^3 \subset \mathbb{R}^4$ is conformally flat if and only if the following conditions hold</u>

i) $d\alpha \wedge \alpha = d\beta \wedge \beta = d\gamma \wedge \gamma = 0$

ii) $(\mu - \nu)\lambda_\alpha + (\lambda - \nu)\mu_\alpha + (\mu - \lambda)\nu_\alpha = 0$

$(\nu - \lambda)\mu_\beta + (\mu - \lambda)\nu_\beta + (\nu - \mu)\lambda_\beta = 0$

$(\lambda - \mu)\nu_\gamma + (\nu - \mu)\lambda_\gamma + (\lambda - \nu)\mu_\gamma = 0$

<u>Proof.</u> Set $\alpha' = a_1\alpha + b_1\beta + c_1\gamma$,

$\beta' = a_2\alpha + b_2\beta + c_2\gamma$, $\gamma' = a_3\alpha + b_3\beta + c_3\gamma$.

Then $d\alpha \wedge \alpha = 0$ (for instance) if and only if $b_2 + c_3 = 0$, so that condition i) is equivalent to $a_1 = b_2 = c_3 = 0$. Further more, taking i) into account, the Codazzi equation gives

$$\lambda_\beta + a_3(\lambda - \mu) = 0; \; \lambda_\gamma + a_2(\nu - \lambda) = 0$$

and four other equations which are deduced from these two by circular permutation.

In the same way, the conformal flatness condition

$$d\lambda_1 \alpha \wedge \alpha + \lambda_1 d\alpha + \mu_1 \gamma \wedge \beta - \nu_1 \beta' \wedge \gamma = 0$$

or

$$d\lambda_1 \wedge \alpha - 2\nu(\lambda - \mu)\gamma' \wedge \beta + 2\mu(\lambda - \nu)\beta' \wedge \gamma = 0$$

gives

$$(\mu + \nu)\lambda_\beta + (\lambda - \gamma)\mu_\beta + (\lambda - \mu)\nu_\beta + 2a_3\nu(\lambda - \mu) = 0$$

$$(\mu + \nu)\lambda_\gamma + (\lambda - \nu)\mu_\gamma + (\lambda - \mu)\nu_\gamma - 2a_2\mu(\lambda - \nu) = 0$$

Using the Codazzi equations in the form obtained above, this proves our claim: After condition i) has been taken into account, there are six equations to check, the two just above and the four which are deduced from those two by circular permutation. But the Codazzi equations force these equations to be equivalent to three of them.

□

A beautiful, but mysterious interpretation of these conditions has been given by Elie Cartan . For each tangent plane to M, consider the homogeneous cone whose equation is

$$\lambda\alpha^2 + \mu\beta^2 + \nu\gamma^2 = 0$$

Using classical arguments about pencils of conics (cf. [Br] for instance), we see that there exist 6 planes directions (two real, four pair-wise complex conjugate) which give circular sections. Indeed, there are three values of t for which the quadratic form

$$\lambda\alpha^2 + \mu\beta^2 + \nu\gamma^2 - t(\alpha^2 + \beta^2 + \gamma^2)$$

is degenerate $(t = \lambda, \mu, \nu$ of course!). It is then decomposed into a product of linear forms, and these linear forms give the plane dirctions we want. If for instance $\lambda > \mu > \nu$, they are

$$\sqrt{\lambda-\mu}\ \ \beta \pm i\sqrt{\lambda-\nu}\ \ \gamma = 0$$

$$\sqrt{\mu-\nu}\ \ \gamma \pm \sqrt{\lambda-\mu}\ \ \alpha = 0$$

$$\sqrt{\lambda-\nu}\ \ \alpha \pm i\sqrt{\mu-\nu}\ \ \beta = 0$$

Let us call these planes the <u>umbilical planes.</u>

4. <u>Theorem.</u> <u>A generic</u> $M^3 \subset \mathbb{R}^4$ <u>is conformally flat if and only if the umbilical distributions are integrable</u>.

(Of course, for complex planes this makes sense only formally).

<u>Proof.</u> Taking the exterior derivative of $\omega = \sqrt{\lambda-\mu}\,\beta + \varepsilon i\sqrt{\lambda-\nu}\,\gamma$ (where $\varepsilon = \pm 1$) gives

$$d\omega = \sqrt{\lambda-\mu}\ d\beta + \varepsilon i\ \sqrt{\lambda-\nu}\ \ d\gamma$$

$$+ \frac{(d\lambda - d\mu)}{2\sqrt{\lambda-\mu}} \wedge \beta + \varepsilon\, \frac{i(d\lambda - d\nu)\wedge\gamma}{2\sqrt{\lambda-\upsilon}}$$

Then $d\omega\wedge\omega = (\lambda-\mu)d\beta\wedge\beta + (\nu-\lambda)d\gamma\wedge\gamma$

$$+ \varepsilon i\Big[\sqrt{(\lambda-\mu)(\lambda-\nu)}\ (d\beta\wedge\gamma + d\gamma\wedge\beta) + \frac{1}{2}\sqrt{\frac{\lambda-\nu}{\lambda-\mu}}\ (d\lambda - d\mu)\wedge\beta\wedge\gamma$$

$$+ \frac{1}{2}\sqrt{\frac{\lambda-\mu}{\lambda-\nu}}\ (d\lambda - d\nu)\wedge\gamma\wedge\beta\ \Big]$$

Therefore, the integrability of the umbilical distributions is equivalent to the equations

$$d\alpha\wedge\alpha = d\beta\wedge\beta = d\gamma\wedge\gamma = 0 \quad (\text{since } \lambda,\mu,\nu \text{ are distinct})$$

together with

$$2(\lambda-\mu)(\lambda-\nu)(d\beta\wedge\gamma + d\gamma\wedge\beta) + [(\mu-\nu)d\lambda + (\nu-\lambda)d_{\mu} + (\lambda-\mu)d\nu]\wedge\beta\wedge\gamma = 0$$

and two other conditions obtained by ciruclar permutation. The equivalence of these conditions with i) and ii) of Proposition E.3 is derived using the Codazzi equations in the same way as in the proof of this proposition.

□

F. Some global properties of compact conformally flat manifolds.

Let us begin with a down to earth description of the Pontryagin forms of a Riemannian manifold (M,g). For any local coframe $(e_i)_{1 \leq i \leq m}$ on an open set, the curvature can be written as $R = \Sigma\, \Omega_{ij} \otimes e_i \wedge e_j$, where $\Omega = (\Omega_{ij})$ is an antisymmetric matrix of two forms. The algebra generated by differential forms of even degree is commutative. Therefore, if P is a polynomial on the Lie-algebra $\underline{SO}(n)$ of antisymmetric (n,n) matrices, we can define $P(\Omega)$, a differential form of degree $2k$ if p is homogeneous of degree k.

If we take another orthonormal coframe $(e'_i)_{1 \leq i \leq m}$ on $U' \subset M$, then on $U \cap U'$ we have $e'_i = \sum g_{ij} \cdot e_j$ and $\Omega = g\, \Omega\, g^{-1}$, where $g \in C^\infty(U \cap U', O(n))$. Now, if P is $\mathrm{Ad}\, O(n)$-invariant, $P(\Omega)$ does not depend on the choice of a particular coframe, and all the locally defined $P(\Omega)$ match together to give a differential form $P(R)$ on M.

1. Definition. $P(R)$ is the Pontryagin form associated with the polynomial P.

It can be proved, using the second Bianchi identity, that $P(R)$ is closed. Furthermore, its cohomology class is a differential invariant (cf. [M-S], in particular appendix C), the Pontryagin class associated with P. It can also be proved (cf. [Sp], I) that the algebra of $O(n)$-invariant polynomials over $\underline{SO}(n)$ is generated by the polynomial P_k defined by

$$P_k(X) = \mathrm{tr}(X^{2k})$$

(Clearly $\mathrm{tr}(X^{2k+1}) = 0$). The reader may suspect that this is related with symmetric functions and Newton sums). This fact has the following important consequence.

2. Theorem (Chern-Simons [C-S] , cf. also [Ku]). The Pontryagin forms only depend on the Weyl tensor.

<u>Proof.</u> It is enough to prove it for

$$P_k(R) = \sum_{i,\dots,i_{2/n}} \Omega_{i_1 i_2} \wedge \Omega_{i_2 i_3} \dots \wedge \Omega_{i_{2k} i_1}$$

Now, taking a basis which diagonalizes the Ricci tensor, we have

$$R = \sum_{i<j} W_{ij} \otimes e_i \wedge e_j + 2 \sum_{i<j} (\lambda_i + \lambda_j) e_i \wedge e_j \otimes (e_i \wedge e_j)$$

(cf. B.4). We must prove that we can replace Ω_{ij} by something like $\Omega_{ij} + c_{ij}\, e_i \wedge e_j$. But the first Bianchi identity just says that

$$\sum_j e_j \wedge \Omega_{jk} = 0$$

Therefore we can do this substitution step by step in the formula which gives us $P_k(R)$.

3. <u>Corollary.</u> <u>If a differential manifold carries a conformally flat metric, its Pontryagin classes all vanish.</u>

For an oriented Riemannian manifold with volume form ω , recall that the Hodge operator $* : \Lambda^p M \to \Lambda^{n-p} M$ is defined by

(4) $\quad *\alpha \wedge \beta = g(\alpha,\beta)\omega$

In particular, when n is even, it gives rise to a conformally invariant isomorphism of $\Lambda^{n/2} M$, i.e. we have the following

5. <u>Proposition.</u> <u>For any open set</u> U , <u>the harmonic forms on</u> U <u>only depend on the conformal class of the metric.</u>

This elementary property has important consequences concerning the Weitzenböck formula in middle dimension. Recall (cf. [SБ], n°XVI [Be] for detailed information) that this formula says that the Hodge-de Rham Laplacian $\Delta = d\delta + \delta d$ and the rough Laplacian D^*D (D^* is the formal adjoint of the covariant derivative D) are related by

$$\Delta = D^*D + L_p(R)$$

where $L_p(R)$ is a linear map on the bundle $\Lambda^p M$, which depends linearly on the curvature, and is "universal" in that it depends on p only (for $p = 1$, it is well known that $L_1(R) = \mathrm{Ric}$ as an endomorphism).

6. Theorem (J.P. Bourguignan, [Bo]). *If (M,g) is a Riemannian manifold of dimension $n = 2m$ the Weintzenböck formula for m-forms is given by*

$$\Delta = D^*D + \frac{\mathrm{Scal}}{2n(n-1)} + L'_m(W)$$

(i.e. the order zero part of the formula only depends on scalar and Weyl curvatures).

Proof. Using local existence theorems for elliptic equations, we see that for any $x \in M$ and any $\omega \in \Lambda^m M$, there exists in a neighbourhood of x a harmonic form α such that $\alpha_m = \omega_x$. The preceding propositions and the relation $D^*D\star = \star D^*D$ then prove that $L_m(R)$ commutes with $\star$.

Furthermore, since Δ and D^*D are formally selfadjoint, $L_m(R)$ is symmetric. Now, look at the case $n = 4$. Setting $E = T_xM$, we know that at each point x, $L_m(R)$ belongs to $S^2(\Lambda^2 E)$, which admits the $O(n)$ decomposition (cf. B)

$$S^2(\Lambda^2 E) = \mathbb{R}\, g\cdot g \oplus S^2_0 E \oplus \mathcal{W}E \oplus \mathbb{R}\star$$

(in dim 4, $\Lambda^4 E$ is generated by $\star$). Using Schur's lemma (cf. [N-S]), we see that $L_m(R)$ has not component in $\mathbb{R}\star$, and that its decompositions with respect to the first three irreducible factors is given by

$$L_2(R) = L'_1(\mathrm{Scal}) + L''(\mathrm{Ric} - \frac{\mathrm{Scal}}{4} g) + L'''(W).$$

Now, tensors in $\mathbb{R}\, g \cdot g = \mathbb{R}\, \mathrm{Id}_{\Lambda^2 E}$ clearly commute with $*$. So do tensors in $\mathcal{W}E$, since $\mathcal{W}E$ is generated by $(a \wedge b) \circ (c \wedge d)$ where a,b,c,d are pairwise orthogonal. On the other hand, tensors in $S_0^2 E$ anticommute with $*$: indeed, if $(e_i)_{1 \leq i \leq n}$ is an orthonormal basis, they are generated by $\sum_i (a \wedge e_i) \circ (b \wedge e_i)$ where $g(a,b) = 0$. Therefore $L'' = 0$

The general case $n = 2m$ is proven with the same commutation argument ([Bo], § 8). There is a generalisation of B., namely the decomposition of $S^2(\Lambda^p E)$ into $O(n)$-irreducible components. This is due to R. Kulkarni, cf. [Ku] § 3.

To find the coefficient of Scal in the formula, take the computation for the standard sphere. □

Using the Hodge-de Rham theorem, an immediate consequence is the

7. Theorem (Bourguignon, [Bo]). Let (M,g) be a compact conformally flat Riemannian manifold with positive scalar curvature, of even dimension $2m$. Then $H^m(M,\mathbb{R}) = 0$.

Proof. This is the classical Weintzenböck argument: for $\alpha \in \Omega^m(M)$,

$$\int_M g(\Delta\alpha,\alpha)v_g = \int_M (g(D^*D\alpha,\alpha)v_g + \mathrm{Scal}\; g(\alpha,\alpha))v_g$$

$$= \int_M |D\alpha|^2 v_g + \int_M \mathrm{Scal}\; |\alpha|^2 vg \; .$$

Then if $\Delta\alpha = 0$, $\alpha = 0$.

□

As usual is this situation, with a topological assumption precise information can be caught in the limit case

8. Theorem [L] Let (M,g) be a compact conformally flat Riemannian manifold with zero scalar curvature. Suppose that $\dim M = 2m$, and that $H^m(M,\mathbb{R}) \neq 0$. Then either (M,g) is flat,

or its universal Riemannian cover is the Riemannian product
$S^n \times H^n$.

Proof. Since for m-forms $\Delta\alpha = D^*D\alpha$ in that case, there is a non-zero parallel m-form. This form is invariant under the action of the holonomy group G of M. Now, a by-product of the classification of holonomy groups is the following

Lemma. _If_ (M,g) _is locally irreducible and if_ $G \neq SO(n)$ _and_ $G \neq U(n/2)$, _then_ (M,g) _is Einstein._ (cf. [Be]).

Now, if $E = T_xM$, $\Lambda^m E$ has a one dimensional invariant subspace under the holonomy representation. Then G cannot be $SO(n)$, and cannot be $U(n/2)$ either (unless $n = 4$, but to be in the same time conformally flat and Kähler is so strong that the treatment in that case is easy). Therefore (M,g) is locally reducible (if either it is Einstein, it is flat and already reducible!). Using the de Rham decomposition theorem, this means that (M,g) is locally a Riemannian product. By proposition D.2 , this is possible only if (M,g) is flat or locally isometric to $S^m \times H^m$.

□

REFERENCES

[Br] M. BERGER, Geometrie, tome 4, Cedic-Nathan, to be translated by Springer.

[Be] A. BESSE, Einstein manifolds, to appear in Springer Ergebnisse.

[Bo] J.P. BOURGUIGNON, Les varieties de dimension 4 á courbure harmonique et á signature non nulle sout d'Einstein, Inv. Math. 63 (1981), 263 - 286.

[C] E. CARTAN, La déformation des hypersurfaces dans l'espace conforme á $n \geqq 5$ dimensions, in Oeuvres complétes III.1, 221-286.

[C-D-M] M. DO CARMO, M. DAJCZER and F. MERCURI, Compact conformally flat hypersurfaces, Trans. Amer. Math. Soc. 288 (1985), 189-203.

[C-S] S.S. CHERN and J. SIMONS, Characteristic forms and geometric invariants, Ann. of Math. 99 (1974), 48-69.

[H] R.S. HAMILTON, Three manifolds with positive Ricci curvature, J. Diff. Geom. 17 (1982), 255-306.

[Ku1] R. KULKARNI, On the Bianchi identities, Math. Ann. 199, 175-204 (1972).

[Ku2] R. KULKARNI, Curvature invariants and the index theorem. Presses de l'Université de Montréal.

[L] J. LAFONTAINE, Remarques sur les varietés conformement plates, Math. Ann. 259 (1982), 313-319.

[M-S] J. MILNOR and J. STASHEFF, Characteristic classes, Princeton University Press.

[N-S] M.A. NAIMARK and STERN, Theory of group representations (translated from the Russian), Springer Grundlehren n°246.

[SB] SEMINAIRE ARTHUR BESSE, Geometrie Riemmannienne en dimension 4, Cedic Nathan, Paris.

[SP] M. SPIVAK, Differential Geometry, T.I,II,III and IV.

[W1] H. WEYL, Zur Infinitesimal Geometrie: Einordnung der projectiven und der konformen Auffassung. In Gesammelte Abhandlungen, vol. II, 1968, 195-207.

[W2] H. WEYL, The classical groups, Princeton University Press.

The Theorem of Lelong-Ferrand and Obata

Jacques Lafontaine

Contents

A. Statement of the result

Let M be a compact manifold equipped with a conformal structure, $C(M)$ the conformal group, $C_0(M)$ its neutral component. Then the following spectacular property holds :

1. Theorem i) (Obata, cf. [O2]). If $C_0(M)$ is not compact, M is conformal to the standard sphere.

ii) (Lelong-Ferrand, cf. [L-F]) The same conclusion holds when $C_0(M)$ is replaced by $C(M)$.

We give here a basically self-contained account of i). The proof uses various geometric tricks, which are dispersed in various papers. It has some Riemannian geometric flavour, while the proof of ii) relies on a thorough study of quasi-conformal homeomorphisms. However, it is worth-noting that the degeneracy of conformal homeomorphisms, which occurs when $C(M)$ is not compact (cf. [L-F] § 8) is very similar to what happens with the flow of an "essential conformal vector field" (cf. B.12 below).

In dimension two, the theorem is a classical result on Riemann surfaces: indeed, the Riemann sphere, whose automorphism group is $PSL(2,\mathbb{C})$, is the only compact Riemann surface whose automorphism group is not compact.

The following property is not needed in the sequel, but illustrates the fact that "big conformal groups" are rather rare.

2. Proposition. Any conformal diffeomorphism of $S^p \times H^q$ $(q \geq 1)$ is an isometry.

Proof. For $p = 0$, this is a result of ch. I, § 6 of these notes. Since one can consider $S^p \times H^q$ as the complement of an S^{q-1} in S^{p+q} (cf. ch II, D1), the proof in the general case goes in the same way: Using Liouville's theorem, a conformal diffeomorphism φ of $S^{p+q} - S^{p-1}$ extends itself to a conformal diffeomorphism (that we shall denote by φ again) of S^{p+q}. Now, φ must leave S^{p-1} stable, and this exactly means that $\varphi \in O(p+1) \times O(q,1)$. (The case $p = q = 1$ is a classical result about Riemann surfaces).

Remark. N. Kuiper proved a similar result for complete Einstein manifolds with negative Ricci curvature, cf. [Ku].

B. Some conformal vector fields on S^n

For the standard sphere, such a result is far from true. Indeed, $I(S^n) = O(n+1)$, while $C(S^n) \cong PO(n+1,1)$ has $n+1$ more dimensions and is not even compact. To put some emphasis on this fact, we describe some remarkable non-compact one parameter groups of $C(S^n)$, which will be useful later on.

1. Proposition (folk). Let S^n be the round sphere in $\mathbb{R}^{n+1}$, and f the restriction to S^n of a linear form: namely $f(x) = \langle v,x \rangle$. Then ∇f generates a one parameter non-compact group G of conformal diffeomorphisms, which has two fixed points, one $(p = \frac{V}{|V|})$ attractive and one $(q = \frac{-V}{|V|})$ repulsive. The other orbits of G are open half circles connecting q and p.

Proof. Take $V = (1,0,\dots,0)$, and write explicitely the stereographic projection S of $S^n = \{(X_i)_{0\le i\le n}, \sum |X_i|^2 = 1\}$ onto $\mathbb{R}^n = \{(X_i)_{0\le i\le n}, X_0 = 0\}$, whose coordinates shall be denoted $x_1, x_2, \dots, x_n$. Then $S(X_0, X_1, \dots, X_n) = (x_1, x_n, \dots, x_n)$,

where $x_i = \dfrac{X_i}{1-X_0}$ $(1 \le i \le n)$,

while $X_0 = \dfrac{\Sigma x_i^2 - 1}{\Sigma x_i^2 + 1}$ and $X_i = \dfrac{2x_i}{1+\Sigma x_i^2}$.

Transported to S^n, the one parameter group of homotheties $h_t : x \longmapsto e^t x$ gives $H_t(X) = S^{-1} \circ h_t \circ S(X) = (X_0^t, X_1^t, \dots, X_n^t)$

where
$$X_0^t = \frac{(e^{2t} + 1)X_0 + e^{2t} - 1}{(e^{2t} - 1)X_0 + e^{2t} + 1}$$

$$X_i^t = \frac{2e^t X_i}{(e^{2t} - 1)X_0 + e^{2t} + 1}$$

Taking the order 1 coefficient of the asymptotic expansion of the X_i^t for $t = 0$, we find that the infinitesmal generator of H_t is $\xi(X) = (1 - X_0^2, -X_0X_1, \dots, -X_0X_n) = V - (X,V)X$, which proves all our claims, since $\nabla_{S^n} X_0$ is the projection on $T_X S^n$ of $\nabla_{\mathbb{R}^{n+1}} X_0 = V$.

2. Remark. The above computation can be used to give other such examples. Namely, take the translation group $\tau_t : x \longmapsto (x_1 + t, x_2, \dots, x_n)$. Then the infinitesimal generator of the one parameter group $T_t = S^{-1} \tau_t S$ is

$$\eta(X) = (X_1 - X_1X_0, 1 - X_0 - X_1^2, -X_1X_2, \dots, -X_1X_n)$$

It can be written $\eta = \nabla_{S^n} X_1 + \eta'$, where $\eta'(X) = (X_1, -X_0, 0 \dots, 0)$ is a Killing vector field. There is only one fixed point for T_t , the north pole ,and the other orbits are circles through p .

3. Corollary. The Lie algebra of the Moebius group admits the vector space decomposition $\mathfrak{C} \oplus \mathbb{K}$, where $\mathbb{K}$ is the Lie algebra of Killing vector fields of the sphere, and $\mathfrak{C}$ the vector space

<u>of gradients of linear forms.</u>

<u>Proof.</u> We only have to check the dimensions. But

$$n+1+\frac{n(n+1)}{2} = \frac{(n+1)(n+2)}{2} .$$

For a more detailed description of the Lie algebra structure, see Kobayashi [Ko], ch. IV. For a nice application of these properties to the Nirenberg problem, see Bourguignon-Ezin [B-E].

4. <u>Definition.</u> Let (M,c) be a manifold equipped with a conformal structure. A closed sub-group G of $C(M,c)$ is said to be <u>inessential</u> if there is a metric g defining c such that $G \subset I(M,g)$. We shall say that G is <u>essential</u> if it is not inessential.

5. <u>Examples.</u> i) Since $I(M,g)$ is compact when M is compact, a non-compact closed group of conformal transformations of a compact manifold will always be essential. The group H_t and T_t of the last paragraph give such examples.

ii) On $M = \mathbb{R}^n - \{0\}$, the homothety group $x \to e^t x$ is inessential: it is an isometry group for the metric

$$\frac{1}{r^2} g_{\mathbb{R}^n} = \frac{dr^2}{r^r} + g_{S^{n-1}} .$$

The converse of i) is true. Namely

6. <u>Proposition.</u> <u>A compact subgroup of</u> $C(M,c)$ <u>is inessential.</u>

<u>Proof.</u> Represent c by a Riemannian metric g, and take $\overline{g} = \int_G \gamma^* g \, d\gamma$ ($d\gamma$ is the Haar measure on G). Then it can be checked that $G \subset I(M,\overline{g})$.

□

We can develop the same notions for vector fields.

7. Definition. A vector field X on (M,c) is conformal (some people say Killing-conformal) if its local flow is conformal. (We do not suppose that X is complete).

If g is a metric representing c, and if φ_t is the local flow of X, differentiating $\varphi_t^* g = a_t g$, we get $L_X g = fg$. Since $L_X g$ is twice the covariant derivative of the one-form $X^\flat$ associated with X, taking traces we see that $f = -\frac{1}{n} \operatorname{div} X$. So X is conformal if and only if
$L_X g + \frac{2}{n} (\operatorname{div}_g X) g = 0$.

8. Example. Let $\xi = \nabla f$ be the vector field of proposition B.1. Then

$$L_\xi g = 2Ddf = -\frac{2}{n}(\Delta f) g = -2fg,$$

since f is a first order spherical harmonic.

9. Definition. A conformal vector field on (M,c) is inessential if, for a suitable metric representing c, it becomes a Killing vector field. It is essential if it is not inessential.

10. Example. We have just seen two types of essential vector fields on the sphere. It turns out that all of them belong to one of these types. The proof relies on the

11. Lemma. Let X be an essential conformal vector field on (M,g). Then X admits a zero.

Proof. Suppose that $X \neq 0$ everywhere, and take $\overline{g} = \frac{g}{g(X,X)}$. Then

$$L_X \overline{g} = \frac{1}{g(X,X)} \left(L_X g - \frac{(X \,.\, g(X,X))}{g(X,X)} \, g \right)$$

On one hand, $L_X g = -\frac{2}{n}(\operatorname{div} X) g$, since X is conformal. On the other hand, for any vector field X,

$$X \,.\, g(X,X) = 2(D_X X, X) = (L_X g)(X,X)$$

so that $X \cdot g(X,X) = -\frac{2}{n}(\operatorname{div} X) g(X,X)$, using conformality again. Therefore $L_X \bar{g} = 0$. □

12. Theorem. *Let X be an essential conformal vector field on the standard sphere, and f_t the one parameter group it generates. Then either X has exactly one fixed point p_0, and for any other point $\lim_{t\to\pm\infty} f_t(p) = p_0$; or X has exactly two fixed points p_+ and p_-, and for any other point p, $\lim_{t\to+\infty} f_t(p) = p_+$, $\lim_{t\to-\infty} f_t(p) = p_-$.*

Proof. Using the lemma, X has at least a fixed point p. Using a stereographic projection with pole p, one gets a one parameter group of conformal transformations of $\mathbb{R}^n$, i.e. a one parameter group of similarities (cf. ch. I). By a Cauchy type argument, any non isometric similarity has a unique fixed point, so there are two possibilities:

a) either f_t gives rise to an isometry group of $\mathbb{R}^n$. Then this group has no fixed point on $\mathbb{R}^n$ (otherwise it would be compact), and X would not be essential. (Proposition B.6). This is our first case.

b) or f_t contains a non-isometric similarity. Since the group f_t is commutative, it is a one parameter group of similarities having a common fixed point q. There is no other fixed point, otherwise f_t would reduce to the identity. This is our second case, p_- is the stereographic projection of q and $p = p_+$.

Remark. This proof is different from Obata's (cf. [O1]), who uses the Lie algebra structure of the conformal group.

We shall also need the following "infinitesimal version" of Liouville theorem.

13. Proposition. *Let $U \subset S^n$ an open set, and X a conformal vector field on U. If $n > 2$, X is the restriction to U of a conformal vector field on S^n.*

Proof. We can apply Liouville's theorem to the local flow of X.

□

Remark. A byproduct of this result is that every conformal vector field on S^n is complete. This is false for other constant curvature spaces. For example, the vector field $X(x) = |x|^2 e_1 - \langle e_i, x\rangle x$ is not complete on $\mathbb{R}^n$. Indeed, its flow is $I \circ T_t \circ I$, where T_t is the group of translations $x \longmapsto x + te_1$ and $I(x) = \frac{x}{\langle x,x\rangle}$: it is not globally defined on $\mathbb{R}^n$. On H^n, such examples are still easier to get: in the Poincaré model $\frac{\Sigma dx_i^2}{(1-|x|^2)^2}$, take $X(x) = \sum x^i \frac{\partial}{\partial x^i}$.

C. The conformal flatness

Before we begin with the proof itself, let us collect some basic facts about conformal and isometry groups.

Fact 1. $C(M,g)$ is a Lie group. If $M' \subset M$ is an open subset of M which is invariant under $C(M,g)$, then $C(M,g)$ acts effectively on $\left(M', g_{|M'}\right)$ as a closed subgroup of $C(M', g_{|M'})$. See [Ko], ch. 1, for the first part, and [O2] for the second (or prove it directly, by using the conformal frames of ch. I).

Fact 2. A closed subgroup of $I(M,g)$ is compact if and only if there is a point $p \in M$ such that the orbit through p is compact. This is an exercise on general topology.

Fact 3. (Montgomery and Zippin, [M-Z]). Any non-compact connected Lie group contains a closed one parameter subgroup isomorphic to $\mathbb{R}$.

The first step of the proof is purely Riemannian.

1. Theorem. *Let (M,g) be a compact Riemannian manifold such that the group $C_0(M,g)$ is essential. Then (M,g) is conformally flat.*

<u>Proof.</u> Suppose first $\dim M \geq 4$. Let $N = \{p \in M, W_p \neq 0\}$, and let N_0 a connected component of N. It is again an open set. By fact 1 and connectedness, $C_0(M,g)$ is a closed subgroup of $C_0(N_0,g)$. Now, we have $C_0(N,g) = I_0(N_0,\overline{g})$ for the metric $\overline{g} = \sqrt{g(W,W)}\; g$, using ch. II, C.3.

Now, using fact 3, $C_0(M,g)$ contains a closed one parameter subgroup G isomorphic to $\mathbb{R}$, generated by an (essential) conformal vector field X. The group G is also closed in $C_0(N_0,g) = I_0(N_0,\overline{g})$, therefore, using fact 2, the orbit $G(p)$ of each point of N_0 is non-compact. In particular, X is nowhere zero on N_0.

Consider $T = X \otimes X \otimes X \otimes X \otimes W$. It is a G-invariant tensor, as a tensor product of G-invariant tensors. Since it is of type $(4,4)$, $\|T\|$ is also G-invariant.

Let $p \in N_0$, and $q \in \overline{G(p)}$ (adherence of the orbit of p <u>in M</u>). As $\|T\|$ is a non zero constant on $G(p)$, $\|T\|_q \neq 0$. Therefore $q \in N_0$, and since $G(p)$ is closed in N_0, $q \in G(p)$. So $G(p)$ is closed in M, which is compact: $G(p)$ is compact, a contradiction. So N_0 is empty, and $W \equiv 0$. Then use the Weyl-Schouten theorem (cf. II, C.9).

The same proof works in dimension 3, with minor changes. If h is the Schouten tensor, a tedious but straightforward computation proves that $d^D h$ is conformally invariant. In the first part of the proof, take $\overline{g} = [g(d^D h, d^D h)]^{2/3} g$, in the second part take $T = X \otimes X \otimes X \otimes d^D h$. The same argument proves that $d^D h \equiv 0$.

D. <u>Last step of the proof: using the developing map.</u>

1. <u>Theorem.</u> <u>Let</u> (M,g) <u>be a conformally flat manifold such that</u> $C_0(M,g)$ <u>is essential. Then</u> (M,g) <u>is globally conformal to the standard sphere or to</u> $S^n - \{p_0\} = \mathbb{R}^n$.

2. <u>Lemma.</u> <u>Let</u> (M,g) <u>be a Riemannian manifold</u> $(\widetilde{M},\widetilde{g})$ <u>a</u>

<u>Riemannian covering and</u> $p:\tilde{M} \to M$ <u>the projection. Then</u> $C_0(M,g)$ <u>acts on</u> $\tilde{M}$ <u>as a closed subgroup of</u> $C_0(\tilde{M},\tilde{g})$. <u>Furthermore, any closed one parameter group</u> G <u>which is essential on</u> M <u>is essential on</u> $\tilde{M}$.

<u>Proof.</u> The first part comes from the covering homotopy theorem, and the closedness claim from the fact that in the conformal group, convergence can be read on the conformal frames, cf. [O2].

Let $G \subset C_0(M,g)$ be a closed essential one parameter group. By lemma B.11, G has a fixed point p, and any point $\tilde{p}$ lying over p is a fixed point for G acting on $\tilde{M}$. If G were not essential on $\tilde{M}$, it would be contained in the isotropy group of $\tilde{p}$ for some metric, and consequently would be compact, a contradiction.

<u>Proof of the theorem.</u> Let $\delta : M \longrightarrow S^n$ a developing map. The main point (which seems to be omitted in [O2]) is to prove that δ is one-one.

Let X be an essential vector field on $\tilde{M}$.

Claim 1. This vector field is projectable on S^n. Indeed, since δ is a local diffeomorphism, there is an open set U and $\tilde{M}$ and a conformal vector field Y on $\delta(U)$ such that $\delta_* X = Y$ on U. Now, Y extends itself to a conformal vector field on S^n (cf. B.13), still denoted by Y. Since everything is analytic (ch. I), we still have $\delta_* X = Y$ on $\tilde{M}$.

Claim 2. $\delta(\tilde{M}) = S^n$ or $S^n - \{p\}$. Y is essential on $\delta(\tilde{M})$, a fortiori on S^n, and its flow must leave $\delta(\tilde{M})$ stable. Then our claim comes from theorem B.12.

Claim 3. δ is injective. (This argument is due to U. Pinkall)

Using the orbit structure of the group generated by Y, we can construct backwards a conformal map φ of $\delta(\widetilde{M})$ into $\widetilde{M}$. There are two cases to consider. a) If $\delta(\widetilde{M}) = S^n$, $\varphi(S^n)$ is an open and closed set of $\widetilde{M}$. Therefore $\widetilde{M} = \varphi(S^n)$ is compact, and δ is a diffeomorphism.

b) If $\delta(\widetilde{M}) = S^n - \{p\} = \mathbb{R}^n$, φ is still surjective. Otherwise, taking the developing map in the neighbourhood of a boundary point q of $\varphi(\widetilde{M})$, we can extend φ across q, a contradiction. The map $\delta \circ \varphi$ is an everywhere defined conformal map of $\mathbb{R}^n$: it is a global similarity, and therefore a diffeomorphism. Since φ is surjective, δ is injective. Therefore, M is conformally covered by S^n or $S^n - \{p\}$. Let X be the essential conformal vector field we started with. The fixed points of X on $\widetilde{M}$ must cover the fixed points of X on M. So we have a covering of order two at most, and $S^n - \{p\}$ is ruled out. We cannot have a covering of order two by S^n either: in that case the vector field X on S^n should have two fixed points, but since one of them is attractive and the other repulsive, they cannot be identified when going to the quotient. Therefore $M = S^n$.

□

REFERENCES

[B-E] J.P. BOURGUIGNON and J.-P. EZIN, Scalar curvature functions in a conformal class of metrics and conformal transformations, to appear.

[Ku] N. KUIPER, Einstein spaces and connections, Proc. Amsterdam III (1950), 1560-76.

[Ko] S. KOBAYASHI, Transformation groups in differential geometry, Springer Ergebnisse 70.

[LF] J. LELONG-FERRAND, Transformations conformes et quasi-conformes des varietés riemanniennes, Acad. Roy. Belgique Sci. Mem. Coll. 8 (2), 39 (1971).

[M-Z] J. MONTGOMERY and L. ZIPPIN, Tranformation groups.

[O1] M. OBATA, Conformal transformations of Riemannian manifolds, J. Diff. Geom. 4 (1970), 311-333.

[O2] M. OBATA, The conjectures about conformal transformations, J. Diff. Geom. 6 (1971), 247-258.

Conformal Transformations between Einstein Spaces

Wolfgang Kühnel

Contents

This contribution is concerned with the following question which has already been studied by H.W. Brinkmann [Br 2] in 1925: "When can an Einstein space be mapped conformally on some (possibly different) Einstein space and in how many ways can it be so mapped?" Brinkmann was able to answer this question completely in terms of local coordinates. The discussion of the corresponding global problem began much later, and it seems that so far no complete answer to the following question has been given (not even in A. Besse's new book [Be]):

<u>When can a complete Einstein space be mapped conformally on some (possibly non-complete) Einstein space?</u>

Answers have been given under the assumption that both spaces are complete (see N.H. Kuiper [Kui 2], K. Yano and T. Nagano [Y-N] and R. Kulkarni [Kul]), and some generalizations have been obtained for spaces with parallel Ricci tensor or only with constant scalar curvature and additional restrictions. On the other hand it seems that Brinkmann's results have been forgotten for a while and partially reinvented by several authors. Moreover it even happened that theorems have been published which contradict Brinkmann's results, and independently incorrect answers to the question above have been given (see our discussion of the "standard wrong theorems" at the end of this paper).

For this and other reasons we started to work out all the calculations from the very beginning and to write down these essentially self-contained notes. Our answer to the question above will be given in theorem 27 , and the relations with the various results in various papers will be discussed. We will restrict ourselves to the case of Riemannian manifolds with positive definite metric.

A. Basic formulas

Let (M^n,g) be a Riemannian manifold of dimension n , ∇ the Levi-Civita-connection and

$$R(X,Y)Z = \nabla_X\nabla_Y Z - \nabla_Y\nabla_X Z - \nabla_{[X,Y]}Z$$

the curvature tensor. We write also $\langle X,Y\rangle$ instead of $g(X,Y)$ if this is convenient. The Ricci tensor $\operatorname{trace}(X \to R(X,Y)Z)$ is denoted by $\operatorname{Ric}(Y,Z)$, and the normalized scalar curvature by $\rho = \frac{1}{n(n-1)}\operatorname{trace}\operatorname{Ric}$. The trace itself is non-normalized such that $\operatorname{trace} g = n$. An Einstein space is a Riemannian manifold of dimension $n \geq 3$ such that the Ricci tensor is a multiple of the metric. It follows that

$$\operatorname{Ric} = (n-1)\cdot\rho\cdot g$$

and that ρ is constant.

We will consider conformal transformations of the metric $\bar{g} := e^{-2\varphi}\cdot g$ with a function $\varphi : M \to \mathbb{R}$. This leads to a Riemannian manifold $(M,\bar{g})$ for which all the quantities above are denoted by $\bar{\nabla}$, $\bar{R}$, $\overline{\operatorname{Ric}}$, $\bar{\rho}$ etc.

It turns out that our problem mentioned in the introduction leads to a simple differential equation for the function $\psi := e^{\varphi}$. The gradient of ψ is denoted by $\nabla\psi$, the Hessian form by $\nabla^2\psi$.

We will also use the Hessian 1-1-tensor H_ψ defined by

$$H_\psi(X) := \nabla_X(\nabla\psi)$$

which implies $\nabla^2\psi(X,Y) = \left\langle \nabla_X(\nabla\psi), Y\right\rangle$. The Laplacian $\Delta\psi$ of ψ is the trace of H_ψ .

1. Lemma. For g and $\bar{g} = \psi^{-2}\cdot g$, $\psi = e^\varphi$ we have the equations:

i) $\bar{\nabla}_X Y = \nabla_X Y - ((X\varphi)Y + (Y\varphi)X - \langle X,Y\rangle\, \nabla\varphi)$

ii)
$$\begin{aligned}\bar{R}(X,Y)Z = R(X,Y)Z &- [\langle X,Z\rangle H_\varphi Y - \langle Y,Z\rangle H_\varphi X] + \\ &+ \left[\nabla^2\varphi(Y,Z) + (Y\varphi)(Z\varphi) - \langle Y,Z\rangle \,\|\nabla\varphi\|^2\right]X - \\ &- \left[\nabla^2\varphi(X,Z) + (X\varphi)(Z\varphi) - \langle X,Z\rangle \,\|\nabla\varphi\|^2\right]Y + \\ &+ \left[(X\varphi)\langle Y,Z\rangle - (Y\varphi)\langle X,Z\rangle\right]\cdot\nabla\varphi \ .\end{aligned}$$

iii) $\overline{\mathrm{Ric}} = \mathrm{Ric} + (\Delta\varphi - (n-2)\|\nabla\varphi\|^2)\cdot g + \frac{n-2}{\psi}\nabla^2\psi$

iv) $\psi^{-2}\,\bar{\rho} = \rho + \frac{2}{n}\Delta\varphi - \frac{n-2}{n}\|\nabla\varphi\|^2 = \rho + \frac{2}{n}\frac{\Delta\psi}{\psi} - \frac{\|\nabla\psi\|^2}{\psi^2}$.

Proof. i) and ii) are standard formulas. They may be found in [Be] or in Lafontaine's contribution [La 2] in this volume.
iii) By a straight forward calculation ii) implies

$$\overline{\mathrm{Ric}}(Y,Z) = \mathrm{Ric}(Y,Z) + (\Delta\varphi - (n-2)\|\nabla\varphi\|^2)\langle Y,Z\rangle + (n-2)(\nabla^2\varphi(Y,Z) + (Y\varphi)(Z\varphi)) \ .$$

The relation

$$\nabla^2 e^\varphi(Y,Z) = e^\varphi(\nabla^2\varphi(Y,Z) + (Y\varphi)(Z\varphi))$$

suggests to use $\nabla^2\psi$ instead of $\nabla^2\varphi$.

iv) follows from iii) by taking the trace.

B. The differential equation associated with the local problem

The problem to be discussed is the following:

Assume that (M,g) *is an Einstein space. Under which conditions there exists a function* $\psi : M \to \mathbb{R}$ *such that* $(M, \bar{g} = \psi^{-2} \cdot g)$ *is again an Einstein space?*

2. Lemma (Brinkmann [Br 2]). *Let* (M,g) *be Einstein,* $\bar{g} = \psi^{-2} \cdot g$. *Then the following conditions are equivalent:*

i) $(M,\bar{g})$ *is Einstein,*

ii) *there exists a function* $\lambda : M \to \mathbb{R}$ *with* $\nabla^2\psi = \lambda \cdot g$,

iii) $\nabla^2\psi = \frac{\Delta\psi}{n} \cdot g$

iv) *there exists a constant* B *such that* $\nabla^2\psi = (-\rho \cdot \psi + B) \cdot g$.

Each of these conditions implies:

v) *there are constants* B,C *such that*

$$\| \nabla\psi \|^2 = -\rho\psi^2 + 2B\psi + C$$

Proof. i) $\Leftrightarrow$ ii) follows directly from lemma 1 iii) .
ii) $\Leftrightarrow$ iii) and iv) $\Rightarrow$ ii) are trivial.
i) $\Rightarrow$ iv) Assume $\nabla^2\psi = \lambda \cdot g$ or equivalently $H_\psi = \lambda \cdot id$ then we conclude the so-called Ricci-indentity

$$R(X,Y)\nabla\psi = \nabla_X(H_\psi Y) - \nabla_Y(H_\psi X) - H_\psi[X,Y]$$

$$= \nabla_X(\lambda Y) - \nabla_Y(\lambda X) - \lambda[X,Y] = (X\lambda)Y - (Y\lambda)\cdot X \quad .$$

It follows $\rho \cdot (Y\psi) = \rho \cdot \langle Y,\nabla\psi\rangle = \frac{1}{n-1} Ric(Y,\nabla\psi) = \frac{1}{n-1} \sum_{i=1}^{n} \langle R(E_i,Y)\nabla\psi,E_i\rangle$

$$= \frac{1}{n-1} \sum_{i=1}^{n} (E_i\lambda \langle Y,E_i\rangle - (Y\lambda)) = -(Y\lambda)$$

where $E_1,\ldots,E_n$ is an ON basis.

Consequently $Y(\rho\psi + \lambda)$ is zero for all Y, hence $B := \rho\psi+\lambda$ is locally constant.

iv) ⇒ v) We calculate the derivative of the function $\|\nabla\psi\|^2 + \rho\psi^2 - 2B\psi$ in an arbitrary direction X.

$$\nabla_X \|\nabla\psi\|^2 + \rho\, \nabla_X \psi^2 - 2B\, \nabla_X\psi = 2\nabla^2\psi(X,\nabla\psi) + 2\rho\, \psi(X\psi) - 2B(X\psi) = 0 .$$

The last equality holds because (iv) just says that $\nabla^2\psi(X,\nabla\psi) = (-\rho\psi + B)(X\psi)$.

3. Proposition. Let (M,g) be a space of constant sectional curvature K and dimension $n \geq 3$ and $\psi : M \to \mathbb{R}$ a function, $\bar{g} := \psi^{-2}g$.

Then the following conditions are equivalent:

i) $(M,\bar{g})$ is a space of constant sectional curvature $\bar{K}$,

ii) $(M,\bar{g})$ is Einstein,

iii) $\nabla^2\psi = \frac{\Delta\psi}{n}\cdot g$

iv) $\nabla^2\psi = (-K\psi + B)\cdot g$ for a constant B.

Each condition implies that $K,\bar{K}$ satisfy $\bar{K} = \psi^2 K + \frac{2}{n}\cdot\psi\Delta\psi - \|\nabla\psi\|^2$.

In the case $n = 2$ i) ⇔ iii) ⇔ iv) is true (compare prop. 4).

Proof. By assumption $K = \rho$ is constant.

i) ⇒ ii) is trivial, and ii) ⇔ iii) ⇔ iv) hold by lemma 2.

It remains to show iii) ⇔ i). Let σ the 2-plane spanned by two vectors X,Y which are orthonormal with respect to g.

Let $\bar{K}_\sigma$ denote the sectional curvature of σ in $(M,\bar{g})$.

Then 1(ii) implies

$$\begin{aligned}\psi^{-2}\,\overline{K_\sigma} &= \langle\bar{R}(X,Y)Y,X\rangle \\ &= K + \nabla^2\varphi(Y,Y) + \nabla^2\varphi(X,X) + (Y\varphi)^2 + (X\varphi)^2 - \|\nabla\varphi\|^2 \\ &= K + \frac{2}{n}\cdot\frac{\Delta\psi}{\psi} - \|\nabla\varphi\|^2 .\end{aligned}$$

Therefore $\bar{K}_\sigma$ is a function on M not depending on σ. By Schur's theorem $\bar{K}$ is constant.

4. <u>Proposition.</u> <u>Let</u> (M,g) <u>be of constant scalar curvature</u> ρ <u>and</u> $\psi : M \to \mathbb{R}$ <u>be a function satisfying</u> $\nabla^2\psi = \frac{\Delta\psi}{n} g$. <u>Then</u> i) <u>and</u> ii) <u>are equivalent.</u>

i) $(M,\psi^{-2}g)$ <u>has constant scalar curvature,</u>
ii) <u>there is a constant</u> B' <u>such that</u> $\Delta\psi = -n\cdot\rho\cdot\psi + B'$.

<u>Proof.</u> For the scalar curvature $\bar{\rho}$ of $\bar{g} = \psi^{-2}g$ we have 1(iv)

$$\bar{\rho} = \psi^2(\rho + \frac{2}{n}\Delta\varphi - \frac{n-2}{n}\|\nabla\varphi\|^2) = \psi^2\rho + \frac{2}{n}\psi\cdot\Delta\psi - \|\nabla\psi\|^2 .$$

It follows that for arbitrary X

$$X(\bar{\rho}) = 2\rho\ \psi(X\psi) + \frac{2}{n}(X\psi)\Delta\psi + \frac{2}{n}\psi\ X(\Delta\psi) - 2\nabla^2\psi(X,\nabla\psi)$$

$$= 2\psi\left[\rho(X\psi) + \frac{1}{n}X(\Delta\psi)\right] .$$

Consequently the constancy of $\bar{\rho}$ is equivalent to the constancy of $n\rho\psi + \Delta\psi$.

C. Concircular mappings

A concircular mapping is a conformal transformation preserving geodesic circles (for a definition see below). This is a priori not related with our problem concerning conformal transformations between Einstein spaces. On the other hand a posteriori it turns out to be the same problem because the corresponding differential equation is exactly the same. The concept of concircular mappings has been introduced by K. Yano [Y 1] and independently by A. Fialkow [Fi] in the more general context of "conformal geodesics".

Let $c : [0,L] \to (M,g)$ be a smooth curve, parametrized by arc length. We denote its derivatives by $\dot{c}$, $\ddot{c} = \nabla_{\dot{c}}\dot{c}$, $\dddot{c} = \nabla_{\dot{c}}\nabla_{\dot{c}}\dot{c}$ etc. If these derivatives are linearly independent we can define the

<u>Frenet frame</u> $e_1,\dots,e_k$ and the <u>geodesic Frenet curvatures</u> $\kappa_1,\dots,\kappa_{k-1}$ in the usual way:

$$\begin{aligned}\dot{e}_1 &= \kappa_1 e_2 \\ \dot{e}_i &= -\kappa_{i-1}e_{i-1} + \kappa_i\, e_{i+1}\ , \ i = 2,\dots,k-1 \ .\end{aligned}$$

c is called <u>geodesic</u> if $\kappa_1 = 0$, and it is called a <u>geodesic circle</u> if κ_1 is constant and $\kappa_2 = 0$ (a geodesic is also a geodesic circle). If $\kappa_1 \neq 0$ for such a geodesic circle then we have the beginning e_1,e_2 of the Frenet frame. $\kappa_2 = 0$ then means that $\dot{e}_2$ lies in the (e_1,e_2)-plane. In the following we write κ instead of κ_1 .

Examples of geodesic circles are small circles on the sphere. It is not required that a geodesic circle is a closed curve. It might be something like a spiral even if the length is infinite.

A diffeomorphism $f : (M,g) \to (\bar{M},\bar{g})$ between two Riemannian manifolds is called <u>concircular</u> if it maps geodesic circles in M to geodesic circles in $\bar{M}$. The original definition by K. Yano [Y 1] required a priori that f is conformal. We will see in 7 below that f is indeed necessarily conformal.

5. <u>Lemma</u> (Yano [Y 1]): <u>c is a geodesic circle if and only if $\dddot{c}$ is a scalar multiple of $\dot{c}$. In this case necessarily</u> $\dddot{c}=-\langle\ddot{c},\ddot{c}\rangle\cdot\dot{c}$.

<u>Proof.</u> If c is a geodesic circle then κ is a constant and the Frenet equations read as

$$\begin{aligned}\ddot{c} = \dot{e}_1 &= \kappa e_2 \\ \dot{e}_2 &= -\kappa e_1 \ .\end{aligned}$$

It follows $\dddot{c} = (\kappa e_2)^{\cdot} = -\kappa^2\dot{c}$.

Vice versa in general

$$\dddot{c} = (\ddot{c})^{\cdot} = \kappa\dot{e}_2 + \dot{\kappa}e_2 = -\kappa^2 e_1 + \dot{\kappa}e_2 \ .$$

If $\ddot{c}$ is a multiple of $e_1 = \dot{c}$ then $\dddot{c} = -\kappa^2 e_1$ and $\dot{\kappa} = 0$, i.e. κ is constant.

If $c = c(t)$ is parametrized by an arbitrary parameter, we denote its derivatives by c', $c'' = \nabla_{c'} c'$, $c''' = \nabla_{c'} \nabla_{c'} c'$ etc. Regarding $c = c(s(t))$ we obtain the following relations:

$$c' = \|c'\| \cdot \dot{c}$$

$$c'' = \|c'\|^2 \cdot \ddot{c} + \frac{<c',c''>}{\|c'\|} \cdot \dot{c}$$

$$c''' = \|c'\|^3 \cdot \dddot{c} + 3<c',c''>\ddot{c} + \left(\frac{<c',c''>}{\|c'\|}\right)' \cdot \dot{c}$$

6. <u>Corollary.</u> c <u>is a geodesic circle if and only if</u> $c''' - 3\frac{<c',c''>}{\|c'\|^2} c''$ <u>is a multiple of</u> c' (<u>or of</u> $\dot{c}$).

<u>Proof.</u> The equation above imply directly

$$c''' - 3\frac{<c',c''>}{\|c'\|^2} c'' + 3\frac{<c',c''>^2}{\|c'\|^3}\dot{c} = c''' - 3<c',c''>\ddot{c}$$

$$= \|c'\|^3 \dddot{c} + \left(\frac{<c',c''>}{\|c'\|}\right)' \dot{c} \quad .$$

Hence $\dddot{c}$ is a multiple of $\dot{c}$ if and only if $c''' - 3\frac{<c',c''>}{\|c'\|^2} c''$ is a multiple of $\dot{c}$.

7. <u>Corollary</u> (Vogel [Vo 1]). <u>Every concircular diffeomorphism is necessarily conformal.</u>

<u>Proof.</u> Without loss of generality we consider a curve $c(s)$ in M endowed with two metrics $g, \bar{g}$ where s is the arc length with respect to g . For $\bar{g}$ this parameter s is not distinguished and therefore we have to compare the various derivatives:

$$\dot{c} \;,\; \ddot{c} = \nabla_{\dot{c}}\dot{c} \;,\; \dddot{c} = \nabla_{\dot{c}}\nabla_{\dot{c}}\dot{c} \quad ,$$

$$c' = \dot{c} \;,\; c'' = \bar{\nabla}_{\dot{c}}\dot{c},\; c''' = \bar{\nabla}_{\dot{c}}\bar{\nabla}_{\dot{c}}\dot{c} \quad .$$

The first observation is that $c'' - \ddot{c} = \bar{\nabla}_{\dot{c}}\dot{c} - \nabla_{\dot{c}}\dot{c}$ is a tensor, i.e. its value at a point depends only on $\dot{c}$ at this point.

Let us fix a point $p \in M$ and two orthonormal vectors X, Y at p. There exists a family c_κ, $\kappa > 0$ of geodesic circles c_κ through p with respect to g such that κ is the constant curvature of c_κ and

$$\dot{c}_\kappa|_p = X$$

$$\ddot{c}_\kappa|_p = \kappa \cdot Y \ .$$

It follows that $c''_\kappa - \ddot{c}_\kappa$ is bounded independently of κ, i.e. $\lim_{\kappa\to\infty} \frac{1}{\kappa} c''_\kappa|_p = \lim_{\kappa\to\infty} \frac{1}{\kappa} \ddot{c}_\kappa|_p = Y$.

Similarly $c'''_\kappa - \dddot{c}_\kappa$ is bounded by a constant times κ, i.e. $\lim_{\kappa\to\infty} \frac{1}{\kappa^2} c'''_\kappa|_p = \lim_{\kappa\to\infty} \frac{1}{\kappa^2} \dddot{c}_\kappa|_p = -X$,

where the last equality follows from lemma 5. $\dddot{c}_\kappa$ is a multiple of $\dot{c}_\kappa$ and lemma 6 tells us that

$$c''' - 3\,\frac{\bar{g}(c'_\kappa, c''_\kappa)}{\bar{g}(c'_\kappa, c'_\kappa)} \cdot c''_\kappa$$

is also a multiple of $\dot{c}_\kappa$ where $\dot{c}_\kappa|_p = X$.

On the other hand

$$\lim_{\kappa\to\infty} \frac{1}{\kappa^2}\left(\dddot{c}_\kappa - c'''_\kappa + 3\,\frac{\bar{g}(c'_\kappa, c''_\kappa)}{\bar{g}(c'_\kappa, c'_\kappa)} \cdot c''_\kappa\right)\Big|_p$$

$$= \lim_{\kappa\to\infty} \frac{1}{\kappa^2}(\dddot{c}_\kappa - c'''_\kappa)|_p + 3 \lim_{\kappa\to\infty} \frac{\bar{g}(c'_\kappa, \frac{1}{\kappa}c''_\kappa)}{\bar{g}(c'_\kappa, c'_\kappa)} \cdot \frac{1}{\kappa} c''_\kappa|_p$$

$$= 0 + 3\,\frac{\bar{g}(X,Y)}{\bar{g}(X,X)} \cdot Y$$

which is impossible unless $\bar{g}(X,Y) = 0$. Because p, X, Y have been arbitrary it follows that the transformation $g \to \bar{g}$ preserves orthogonality and is therefore conformal.

8. <u>Proposition.</u> <u>A conformal transformation</u> $g \to \bar{g} = \psi^{-2} \cdot g$ <u>is concircular if and only if there is a scalar function</u> λ <u>satisfying</u> $\nabla^2\psi = \lambda \cdot g$. <u>It follows in this case that</u> $n \cdot \lambda = \Delta\psi$.

<u>Proof.</u> The transformation $g \to \bar{g}$ is concircular if and only if for any curve parametrized by g-arclength the following conditions are equivalent:

(i) $\dddot{c}$ is a multiple of $\dot{c}$

(ii) $c''' - 3\,\dfrac{\bar{g}(c',c'')}{\bar{g}(c',c')}\,c''$ is a multiple of $\dot{c}$.

We calculate c'' and c''' in terms of $\dot{c},\ddot{c},\dddot{c}$ using lemma 1 and the notation $\psi = e^{\varphi}$.

$$c'' = \bar{\nabla}_{\dot{c}}\, \dot{c} = \ddot{c} - 2\dot{\varphi}\, \dot{c} + \nabla\varphi$$

$$c''' = \bar{\nabla}_{\dot{c}}\, \bar{\nabla}_{\dot{c}}\dot{c} = \dddot{c} - 3\dot{\varphi}\, \ddot{c} + \left(4\dot{\varphi}^2 - 3\ddot{\varphi} + \nabla^2\varphi(\dot{c},\dot{c}) + \|\nabla\varphi\|^2\right)\cdot\dot{c} - 2\dot{\varphi}\nabla\varphi + \nabla_{\dot{c}}\nabla\varphi.$$

Therefore the following equation holds with a certain scalar factor α :

$$c''' - 3\,\frac{\bar{g}(c',c'')}{\bar{g}(c',c')}\,c'' = c''' - 3\,\frac{g(\dot{c},c'')}{g(\dot{c},\dot{c})}\,c''$$

$$= \dddot{c} - 3\dot{\varphi}\, \ddot{c} - 2\dot{\varphi}\nabla\varphi + \nabla_{\dot{c}}\nabla\varphi + 3\dot{\varphi}(\ddot{c} + \nabla\varphi) + \alpha\dot{c}$$

$$= \dddot{c} + \dot{\varphi}\nabla\varphi + \nabla_{\dot{c}}\,\nabla\varphi + \alpha\dot{c}\ .$$

Thus if (i) and (ii) are equivalent then

$$\frac{1}{\psi}\,\nabla_{\dot{c}}\nabla\psi = \dot{\varphi}\,\nabla\varphi + \nabla_{\dot{c}}\nabla\varphi$$

must be a multiple of $\dot{c}$ and vice versa. Because there are geodesic circles with arbitrarily given tangent $X = \dot{c}$ it follows

$$\nabla_X\, \nabla\psi = \lambda(X)\cdot X$$

for all tangent vectors X . For a fixed point the expression $\nabla_X\, \nabla\psi$ is linear in X and therefore λ depends only on the point and not on X . Regarding $\lambda : M \to \mathbb{R}$ as a function we have

$\nabla_X \nabla\psi = \lambda\cdot X$ or equivalently $\nabla^2\psi = \lambda\cdot g$.

9. Corollary. For a conformal transformation which maps geodesics to geodesics the conformal factor is constant (such maps are called homotheties).

Proof. By assumption $\nabla_{\dot c}\dot c = 0$ and $\bar\nabla_{\dot{\bar c}}\dot{\bar c} = 0$ are equivalent where $\dot{\bar c} = e^{\varphi}\dot c$ is the unit tangent with respect to $\bar g$. As in the proof of 8 we calculate

$$\bar\nabla_{\dot{\bar c}}\dot{\bar c} = e^{\varphi}\ \bar\nabla_{\dot c}(e^{\varphi}\dot c)$$

$$= e^{\varphi}\ (\dot\varphi\ e^{\varphi}\ \dot c + e^{\varphi}\ \bar\nabla_{\dot c}\dot c)$$

$$= e^{2\varphi}(\dot\varphi\ \dot c + \ddot c - 2\dot\varphi\ \dot c + \nabla\varphi) \quad .$$

If $\ddot c$ and $\bar\nabla_{\dot{\bar c}}\dot{\bar c}$ both are zero then

$$\langle\dot c, \nabla\varphi\rangle\dot c = \dot\varphi\dot c = \nabla\varphi$$

for an arbitrary vector $X = \dot c$ which is impossible unless $\nabla\varphi = 0$.

10. Corollary (Yano [Y 1]). Let (M,g) be an Einstein space [or space of constant sectional curvature] and $g \to \bar g = \psi^{-2} g$ a conformal transformation. Then (i) and (ii) are equivalent:

(i) $(M,\bar g)$ is an Einstein space [or space of constant sectional curvature]

(ii) The transformation $g \to \bar g$ is concircular.

This follows directly from 2,3 and 8 .

D. Local solutions of the differential equation $\nabla^2\psi = \lambda\cdot g$

In section B and C we have seen how the differential equation comes in which requires that $\nabla^2\psi$ is a multiple of the metric. Because every constant function is a trivial solution we will assume in the following that (M,g) admits a (locally) non-constant function $\psi : M \to \mathbb{R}$ satisfying

$$(*) \qquad \nabla^2\psi = \frac{\Delta\psi}{n}\cdot g \quad .$$

It turns out that the geometry of such a manifold is of particularly simple type. Immediate consequences of (*) are the following observations:

11. Lemma. <u>Let $U \subseteq M$ be an open set without critical points of ψ, i.e. $\nabla\psi|_p \neq 0$ for all $p \in U$. Then the following holds:</u>

(i) <u>the trajections of $\nabla\psi$ are geodesics (up to parametrization)</u>

(ii) <u>the level hypersurfaces of ψ are totally umbilical,</u>

(iii) <u>$\|\nabla\psi\|$ is constant along the levels of ψ.</u>

(iv) <u>there exists a function α of ψ such that $\nabla\Delta\psi = \alpha\nabla\psi$,</u>

(v) <u>the sectional curvature of every plane containing $\nabla\psi$ is</u> $K = -\frac{\alpha}{n}$.

<u>Proof.</u> The unit normal of a level hypersurface $\{x \in M | \psi(x) = c\}$ is $N = \frac{\nabla\psi}{\|\nabla\psi\|}$. Then the differential equation $\nabla^2\psi = \lambda\cdot g$ reads as follows:

$$\nabla_X(\nabla\psi) = \lambda\cdot X$$

$$\text{or} \quad \nabla_X N = \frac{\lambda}{\|\nabla\psi\|}\cdot\left(X - \frac{\langle X,\nabla\psi\rangle}{\langle\nabla\psi,\nabla\psi\rangle}\nabla\psi\right)$$

for an arbitrary tangent vector X in M .

(i) For $X = N$ we get $\nabla_N N = 0$.

(ii) For X orthogonal to N we compute the shape operator $LX = -\nabla_X N = -\frac{\lambda}{\|\nabla\psi\|}\cdot X$. Thus L is a multiple of the identity.

(iii) For X orthogonal to N we calculate $\nabla_X\|\nabla\psi\|^2 = 2\langle\nabla_X\nabla\psi,\nabla\psi\rangle = 2\frac{\Delta\psi}{n}\langle X,\nabla\psi\rangle = 0$.

(iv) The Ricci identity (see the proof of 2.) implies $0 = \langle R(X,\nabla\psi)\nabla\psi,\nabla\psi\rangle = (X\lambda)\|\nabla\psi\|^2 - ((\nabla\psi)\lambda)\cdot\langle X,\nabla\psi\rangle$, therefore

$X\lambda = 0$ for all X orthogonal to $\nabla\psi$. It follows that $\Delta\psi$ is constant along the levels of ψ, and therefore the gradients of ψ and $\Delta\psi$ are linearly dependent. The scalar factor α in $\nabla\Delta\psi = \alpha\nabla\psi$ is constant along the levels. Hence α may be regarded as a function of ψ.

(v) Let $\langle X,\nabla\psi\rangle = 0$ and $\|X\| = 1$. Then the Ricci identity implies

$$K = \frac{\langle R(X,\nabla\psi)\nabla\psi, X\rangle}{\langle\nabla\psi,\nabla\psi\rangle} = -\frac{\langle\nabla\psi,\nabla\lambda\rangle}{\langle\nabla\psi,\nabla\psi\rangle} = -\frac{\alpha}{n}.$$

In particular this "radial curvature" does not depend on X.

12. Lemma (Brinkmann [Br 2], Fialkow [Fi], Tashiro [T 3]): The following conditions are equivalent:

(i) There exists a nonconstant solution ψ of $\nabla^2\psi = \frac{\Delta\psi}{n}\cdot g$ in a neighborhood of $p \in M$ with $\nabla\psi|_p \neq 0$.

(ii) There exist local coordinates $(u,u_1,\ldots,u_{n-1})$ in a neighborhood of p and a function $\psi = \psi(u)$ with $\psi'|_p \neq 0$ and a $(n-1)$-dimensional Riemannian metric $g_* = g_*(u_1,\ldots,u_{n-1})$ such that

$$g\left(\frac{\partial}{\partial u},\frac{\partial}{\partial u}\right) = 1$$

$$g\left(\frac{\partial}{\partial u},\frac{\partial}{\partial u_i}\right) = 0 \quad \text{for } i = 1,\ldots,n-1$$

$$g\left(\frac{\partial}{\partial u_i},\frac{\partial}{\partial u_j}\right) = \left(\psi'(u)\right)^2\cdot g_*\left(\frac{\partial}{\partial u_i},\frac{\partial}{\partial u_j}\right) \quad \text{for } i,j = 1,\ldots,n-1.$$

(ii) implies that the line element of g may be written as $ds^2 = du^2 + (\psi'(u))^2\cdot ds_*^2$. Such a metric is also called warped product, and g_* is the metric on an ideal level of ψ.

Proof. (ii) $\Rightarrow$ (i) follows by straight forward calculation. u is the arc length parameter on the trajectories of $\nabla\psi = \psi'\cdot\frac{\partial}{\partial u}$ where we write

$$\psi' = \frac{d\psi}{du}, \quad \psi'' = \frac{d^2\psi}{du^2}.$$

Then the claim is $\nabla^2\psi = \psi''\cdot g$.

a) $\nabla_{\frac{\partial}{\partial u}} \nabla\psi = \nabla_{\frac{\partial}{\partial u}} \left(\psi' \frac{\partial}{\partial u}\right) = \psi'' \frac{\partial}{\partial u}$.

b) $\nabla_{\frac{\partial}{\partial u_i}} \nabla\psi = \nabla_{\frac{\partial}{\partial u_i}} \left(\psi' \frac{\partial}{\partial u}\right) = \psi' \nabla_{\frac{\partial}{\partial u_i}} \frac{\partial}{\partial u} = \psi'' \frac{\partial}{\partial u_i}$.

For the last equality one has to calculate certain Christoffel symbols.

(i) ⇒ (ii) Let $c := \psi(p)$ and let $M_c = \{q \mid \psi(q) = c\}$. M_c is a regular level hypersurface of ψ . Choose any coordinate system $u_1,\dots,u_{n-1}$ on M_c and extend this to <u>geodesic parallel coordinates</u> $(u,u_1,\dots,u_{n-1}$ in a neighborhood of p in M . These have the properties

- the u-lines are geodesics with u as arc length.
- $\frac{\partial}{\partial u}$ is orthogonal to every set $\{(u,u_1,\dots,u_{n-1}) \mid u = \text{const.}\}$

expressing the fact that the different u-levels are parallel to each other and that the distance between them is just the difference of the u-values.

On the other hand we have the ψ-levels $\{q \mid \psi(q) = \text{const.}\}$ where the u-level containing p agrees by construction with the ψ-level M_c . Because $\|\nabla\psi\|$ is constant along the ψ-levels (see 11.(iii)) these are also parallel to each other. Hence the u-levels coincide with the ψ-levels and ψ may be regarded as a function of u :

$$\psi = \psi(u) \quad \text{and} \quad \nabla\psi = \psi' \frac{\partial}{\partial u} \ .$$

In these geodesic parallel coordinates $(u,u_1,\dots,u_{n-1})$ we have

$$g\left(\frac{\partial}{\partial u}, \frac{\partial}{\partial u}\right) = 1$$

$$g\left(\frac{\partial}{\partial u_i}, \frac{\partial}{\partial u}\right) = 0 \qquad \text{for } i = 1,\dots,n-1 \ .$$

It remains to show that

$$(\psi'(u))^{-2} \, g_{ij}(u,u_1,\dots,u_{n-1})$$

is independent of u . Then we can introduce this expression as

$g_{*ij}(u_1,\dots,u_{n-1})$ and the assertion follows.

To see this we differentiate

$$\begin{aligned}
\frac{\partial}{\partial u} g_{ij} &= \left\langle \nabla_{\frac{\partial}{\partial u_i}} \frac{\partial}{\partial u}, \frac{\partial}{\partial u_j} \right\rangle + \left\langle \frac{\partial}{\partial u_i}, \nabla_{\frac{\partial}{\partial u_j}} \frac{\partial}{\partial u} \right\rangle \\
&= \left\langle \nabla_{\frac{\partial}{\partial u_i}} \frac{\nabla\psi}{\psi'}, \frac{\partial}{\partial u_j} \right\rangle + \left\langle \frac{\partial}{\partial u_i}, \nabla_{\frac{\partial}{\partial u_j}} \frac{\nabla\psi}{\psi'} \right\rangle \\
&= \frac{2\,\Delta\psi}{n\cdot\psi'} \cdot g_{ij} \\
&= 2\frac{\psi''}{\psi'} \cdot g_{ij}
\end{aligned}$$

where the last equalities follow from the equation (*) .

Therefore for fixed $u_1,\dots,u_{n-1}$ $g_{ij} = g_{ij}(u)$ satisfies the differential equation

$$g_{ij}' = 2\frac{\psi''}{\psi'} g_{ij}$$

or equivalently $\left(\frac{g_{ij}}{\psi'^2}\right)' = \frac{g_{ij}'}{\psi'^2} - 2\frac{\psi''}{\psi'^3} g_{ij} = 0$.

<u>Remark.</u> (i) For surfaces $(n = 2)$ condition (ii) in 12. just says that M is locally isometric with an (abstract) surface of revolution $ds^2 = du^2 + G^2(u)dv^2$.

(ii) Note that we may prescribe the metric g_* on the ideal ψ-level arbitrarily. Any such g_* will lead to a solution of (*). This indicates that in general there is no finite dimensional moduli space for the set of solutions. This is quite different if we are looking for global solutions or if the solution has a critical point (see below).

13. <u>Lemma.</u> <u>For the warped product metric</u> $ds^2 = du^2 + (\psi'(u))^2 ds_*^2$ <u>there hold the following equalities where</u> X,Y,Z <u>always denote vectors orthogonal to</u> $\frac{\partial}{\partial u}$:

(i) $R(X,Y)Z = R_*(X,Y)Z - \frac{\psi''^2}{\psi'^2}\left(\langle Y,Z\rangle X - \langle X,Z\rangle Y\right)$

$R(X,Y)\frac{\partial}{\partial u} = 0$

$R(X,\frac{\partial}{\partial u})\frac{\partial}{\partial u} = -\frac{\psi'''}{\psi'}\cdot X$

(ii) $\mathrm{Ric}(Y,Z) = \frac{1}{\psi'^2}(\mathrm{Ric}_*(Y,Z) - [(n-2)\psi''^2 + \psi'\psi''']\langle Y,Z\rangle)$

$\mathrm{Ric}(Y,\frac{\partial}{\partial u}) = 0$

$\mathrm{Ric}\left(\frac{\partial}{\partial u},\frac{\partial}{\partial u}\right) = -(n-1)\frac{\psi'''}{\psi'}$

(iii) $\psi'^2\cdot\rho = \frac{n-2}{n}\rho_* - \frac{n-2}{n}\psi''^2 - \frac{2}{n}\psi'\psi'''$

(iv) g is Einstein [of constant sectional curvature] $\Leftrightarrow$ g_* is Einstein [of constant sectional curvature] and $\rho = -\frac{\psi'''}{\psi'}$

(if M is 3-dimensional then one has to read: "g_* is of constant Gauss curvature" instead of "g_* is Einstein").

In particular in this case ψ satisfies the equation $\psi'' = -\rho\psi + B$ for some constant $B \in \mathbb{R}$.

Proof. (i) We use the Gauss equation for the level hypersurfaces $u = \text{const.}$ and 11.(ii) :

$$\nabla_X Y = \nabla_{*X} Y + \langle LX,Y\rangle\frac{\partial}{\partial u}$$
$$= \nabla_{*X} Y - \frac{\psi''}{\psi'^2}\langle X,Y\rangle\cdot\nabla\psi\ .$$

This implies

$$\nabla_X\nabla_Y Z = \nabla_{*X}\nabla_{*Y} Z - \frac{\psi''}{\psi'^2}\langle Y,Z\rangle\nabla_X\nabla\psi - \frac{\psi''}{\psi'^2}(\langle X,\nabla_Y Z\rangle + X\langle Y,Z\rangle)\nabla\psi\ .$$

The first assertion follows directly from $\nabla_X\nabla\psi = \psi'' X$. For the second and third one we use $\nabla_X\frac{\partial}{\partial u} = -LX$ and $\nabla_{\frac{\partial}{\partial u}}\frac{\partial}{\partial u} = 0$ from lemma 11 and the Codazzi equation.

(ii) and (iii) follow from (i) by taking the trace.
(iv) follows directly from (ii) for the Einstein case and from (i) for the constant curvature case. The condition $\rho = -\frac{\psi'''}{\psi'}$ or equivalently $\psi'' = -\rho\psi + B$ recovers the same equation from 2(iv). This time it follows from the warped product metric instead of the conformal transformation.

14. Corollary. Let (M,g) be an Einstein space of scalar curvature ρ, and let $p \in M$ be a regular point of a function ψ satisfying $\nabla^2\psi = \frac{\Delta\psi}{n}\cdot g$. Then in a neighborhood of p there are coordinates $(u,u_1,\ldots,u_{n-1})$ such that $\psi = \psi(u)$ and $\psi'' = -\rho\cdot\psi + B$, B being a constant.

This follows from 2. and 12. (compare 13(iv)).
Such a function ψ is called a scalar SC-field in [T 4].

15. Corollary. A 4-dimensional Einstein space admitting a non-constant solution ψ of (*) is of constant sectional curvature.

Proof. By assumption there is a point p with $\nabla\psi|_p \neq 0$. Then 12. and 13.(iv) imply that the metric g on the ideal level is a 3-dimensional Einstein space, hence of constant sectional curvature. Again 13.(iv) implies the assertion.

16. Corollary. A 4-dimensional Einstein space admitting a non-homothetic conformal (or concircular) mapping onto another Einstein space is of constant sectional curvature.

This follows directly from 2., 10. and 15.

The situation is quite different near a critical point of a solution ψ of (*). The origin of the following two lemmata is somewhat mysterious. Lemma 16 appears already in [I-Y]. Then, in a footnote [T 3] p. 255 it is said that the proof "was not exact in some point". We found a complete proof in [T 4] p. 15ff.

17. Lemma (Tashiro [T 4]2.1): Let $p \in M$ be a critical point of a non-constant function ψ satisfying $\nabla^2\psi = \frac{\Delta\psi}{n}\cdot g$. Then there exists a neighborhood U of p such that

(i) p in the only critical point of ψ in U ,

(ii) the level hypersurfaces of ψ coincide in U with the geodesic distance spheres around p . In particular the critical points of ψ are isolated.

Proof. We choose a critical points p as a boundary point of the set of critical points in M . Then every neighborhood of p will contain also regular points. We choose U such that every point in U has a unique and shortest geodesic joining it with p . Let $q \in U$ be a regular point of ψ such that the geodesic joining p and q contains no critical point (except p). From 11.(iii) we know that $\psi(p) \neq \psi(q)$. The level set $A := \{\bar{q} \in U | \psi(\bar{q}) = \psi(q)\}$ contains a point q_o realizing the distance

$$s_o := d(p,q_o) = d(p,A) > 0 \quad .$$

Let γ_o denote the geodesic joining p and q_o . This realizes the distance between $\gamma_o(0) = p$ and $\gamma_o(s_o) = q_o$. By the Gauss lemma γ_o meets A perpendicularly. Consequently by 11.(i) γ_o is the same curve (up to parametrization) as the trajectory of $\nabla\psi$ through q_o .

Any other point $q_1 \in A$ yields similarly a geodesic trajectory γ_1 of $\nabla\psi$. Let $\gamma_1(s_o) = q_1$. Then the claim is that $\gamma_1(0) = p$. To see this let d_M and $d_{A(s)}$ denote the distance function in M and the level $A(s)$ corresponding to the parameter s , respectively. Then for any $s > 0$

$$d_M(\gamma_o(s),\gamma_1(s)) \leq d_{A(s)}(\gamma_o(s), \gamma_1(s))$$

$$= \frac{\psi'(s)}{\psi'(s_o)}\, d_{A(s_o)}(\gamma_o(s_o),\gamma_1(s_o)) \quad .$$

For the last equality we used the warped product metric according to lemma 12. .

It follows

$$\begin{aligned} d_M(\gamma_o(0),\gamma_1(0)) &= \lim_{s\to 0} d_M(\gamma_o(s),\gamma_1(s)) \\ &\leq \lim_{s\to 0} \frac{\psi'(s)}{\psi'(s_o)}\, d_A(\gamma_o(s_o),\gamma_1(s_o)) \\ &= 0 \end{aligned}$$

Therefore $\gamma_1(0) = \gamma_o(0) = p$, and A is contained in the geodesic distance sphere with radius s_o around p. On the other hand it follows that the arc length parameter on the trajectories is just the geodesic distance to p. Therefore p is the only critical point in U, and the ψ-levels coincide there with the geodesic distance spheres around p.

<u>18. Lemma</u> (Tashiro [T 4]2.2). <u>The following conditions are equivalent:</u>

(i) <u>There exists a nonconstant solution</u> ψ <u>of</u> $\nabla^2\psi = \frac{\Delta\psi}{n}\cdot g$ <u>in a neighborhood of</u> $p \in M$ <u>with</u> $\nabla\psi|_p = 0$.

(ii) <u>There exist polar coordinates</u> $(r,u_1,\dots u_{n-1})$ <u>in a neighborhood of</u> p <u>and an even function</u> $\psi = \psi(r)$ <u>with</u> $\psi'(0) = 0$ <u>and</u> $\psi''(0) \neq 0$, <u>such that</u>

$$ds^2 = dr^2 + \frac{(\psi'(r))^2}{(\psi''(0))^2}\cdot ds_1^2$$

<u>where</u> ds_1^2 <u>denotes the line element of the standard unit sphere</u> (S^{n-1},g_1).

<u>Proof.</u> (ii) $\Rightarrow$ (i) follows from lemma 12 for all points except $r = 0$. The evenness of the function $\psi(r)$ (i.e. vanishing of the odd derivatives $\psi'(0),\psi'''(0),\dots$) implies that the right hand side has no proper singularity at $r = 0$. Then by continuity the equation (*) holds also at $r = 0$.

(i) $\Rightarrow$ (ii) It follows from 12. and 17. that we can introduce locally such coordinates $(r,u_1,\dots,u_{n-1})$ such that for $r \neq 0$

$$ds^2 = dr^2 + (\psi'(r))^2\, ds_*^2$$

where ds_*^2 is the line element of the ideal level (M_*, g_*). It remains to show:

a) ψ is an even function and $\psi''(0) \neq 0$,

b) g_* is of constant sectional curvature $(\psi''(0))^2$.

Let X,Y be two orthonormal vectors in M which are tangent to a level hypersurface $M_* = \{q \mid \psi(q) = r_o > 0\}$ for sufficiently small r_o. For the sectional curvatures K_σ and K_σ^* of the (X,Y)-plane in (M,g) and (M_*, g_*) satisfy (lemma 13(i))

$$\begin{aligned} K_\sigma &= g(R(X,Y)Y,X) \\ &= g(R_*(X,Y)Y,X) - \frac{(\psi''(r_o))^2}{(\psi'(r_o))^2} \\ &= \frac{1}{(\psi'(r_o))^2}\left(K_\sigma^* - \left(\psi''(r_o)\right)^2\right) . \end{aligned}$$

On the other hand we know that K_σ^* is independent of r when r tends to zero. Because $\psi'(0) = 0$ (p is a critical point of ψ) it follows that

$$0 = \lim_{r \to 0} (K_\sigma^* - (\psi''(r))^2) = K_\sigma^* - (\psi''(0))^2 .$$

Hence (M_*, g_*) is a space of constant curvature $(\psi''(0))^2$. This must be positive because by 16 we may assume that M_* is a geodesic distance sphere in M_* which is diffeomorphic to S^{n-1}.

Consequently $(\psi''(0))^2 > 0$ and $g_* = \frac{1}{(\psi''(0))^2} \cdot g_1$. By assumption the metric g has no singularity at $r = 0$. This implies that $\psi(r)$ must be an even function and that the equation

$$ds^2 = dr^2 + \frac{(\psi'(r))^2}{(\psi''(0))^2} \cdot ds_1^2$$

is valid for all $r \geq 0$ just as the usual expression of the euclidean metric in polar coordinates.

19. Corollary. Let (M,g) and $(M,\bar{g})$ be two Einstein spaces, $\bar{g} = \psi^{-2} g$, and let $p \in M$ be a critical point of ψ. Then in a neighborhood of p both spaces are of constant sectional curvature.

Proof. We apply 2,3,13 (iv) and 18 .

20. Proposition. Let $(M,g),(M,\bar{g})$, $\bar{g} = \psi^{-2} g$ be given with a nonconstant function ψ.
Then the following conditions are equivalent:

(i) $\bar{R}(X,Y)Z = R(X,Y)Z$ for all X,Y,Z,

(ii) $\overline{\mathrm{Ric}} = \mathrm{Ric}$

(iii) $\nabla^2\psi = \frac{\|\nabla\psi\|^2}{2\psi} g$

(iv) g is a warped product metric,

$$ds^2 = du^2 + (\psi'(u))^2 ds_*^2 ,$$

and there are constants $a,b,c \in \mathbb{R}$ with $b^2 - 4ac = 0$ satisfying $\psi(u) = au^2 + bu + c$.

Moreover each of the conditions implies that ψ has no critical point. If one of the spaces is Einstein then both are Ricci-flat.

Proof. (i) $\Rightarrow$ (ii) is trivial.
To show (ii) $\Rightarrow$ (iii) we use lemma 1 (iii) :

$$\overline{\mathrm{Ric}} - \mathrm{Ric} = \left(\frac{\Delta\psi}{\psi} - (n-1)\frac{\|\nabla\psi\|^2}{\psi^2}\right)\cdot g + (n-2)\frac{\nabla^2\psi}{\psi} .$$

By assumption $\psi \neq 0$ and $\overline{\mathrm{Ric}} = \mathrm{Ric}$, hence $\nabla^2\psi$ must be a multiple of g :

$$\nabla^2\psi = \frac{\Delta\psi}{n}\cdot g$$

and moreover

$$0 = 2\,\frac{n-1}{n}\,\Delta\psi - (n-1)\frac{\|\nabla\psi\|^2}{\psi} .$$

(iii) $\Rightarrow$ (i) can be seen by straightforward calculation using lemma 1(ii). Here it is convenient to write 1(ii) in terms of ψ (compare [La 2]):

$$\psi(\bar{R}(X,Y)Z - R(X,Y)Z) = -\langle X,Z\rangle H_\psi Y + \langle Y,Z\rangle H_\psi X$$

$$+ \nabla^2\psi(Y,Z)\cdot X - \nabla^2\psi(X,Z)\cdot Y$$

$$+ \frac{\|\nabla\psi\|^2}{\psi}(\langle X,Z\rangle Y - \langle Y,Z\rangle X) .$$

Now assume that $p \in M$ is a point with $\nabla\psi|_p = 0$ and that (iii) holds. Then $\nabla^2\psi|_p = 0$ which contradicts lemma 18. Therefore all points are regular, and (iii) $\Leftrightarrow$ (iv) follows from lemma 12. In the coordinates $(u,u_1,\ldots,u_{n-1})$ the equation

$$\nabla^2\psi = \frac{\|\nabla\psi\|^2}{2\psi}\cdot g$$

reads as $\psi'' = \frac{\psi'^2}{2\psi}$ or equivalently $2\psi\psi'' = \psi'^2$. Differentiating again we get $\psi''' = 0$, and there are constants $a,b,c \in \mathbb{R}$ with $\psi(u) = au^2 + bu + c$. The relation $b^2 = 4ac$ follows from $2\psi\psi'' = \psi'^2$.

If (M,g) is Einstein then $\psi'' = -\rho\psi + B$ (see 14). This is possible only if $\rho = 0$, hence $\overline{\mathrm{Ric}} = \mathrm{Ric} = 0$.

E. Examples of global solutions

Example 1 . Spaces of constant sectional curvature

Let $S^n(c^2)$, E^n , $H^n(-c^2)$ denote the simply connected space forms of constant sectional curvature $c^2, 0, -c^2$ respectively. ds_1^2, ds_o^2, ds_1^2 denotes the line element of the metric on $S^n(1)$, E^n, $H^n(-1)$.

a) Sphere $S^n(1)$:

$$ds^2 = du^2 + \sin^2 u \cdot ds_1^2$$

$$ds^2 = du^2 + \cos^2 u \cdot ds_1^2$$

The corresponding solutions of $\nabla^2\psi = \lambda \cdot g$ are the following where c denotes an arbitrary constant:

$$\psi(u) = \cos u + c$$
$$\psi(u) = \sin u + c$$

b) Euclidean space:

$$ds^2 = du^2 + ds_o^2 \quad , \; \psi(u) = u + c$$
$$ds^2 = du^2 + u^2 ds_1^2 \quad , \; \psi(u) = \frac{u^2}{2} + c$$

c) Hyperbolic space $H^n(-1)$:

$$ds^2 = du^2 + \sin h^2 u \; ds_1^2 \quad , \quad (u) = \cos h \; u + c$$
$$ds^2 = du^2 + e^{2u} \; ds_o^1 \quad , \quad (u) = e^u + c$$
$$ds^2 = du^2 + \cos h^2 u \; ds_{-1}^2 \quad , \quad (u) = \sin h \; u + c$$

All these solutions of $\nabla^2\psi = \lambda \cdot g$ are globally defined. However, only such solutions can be global conformal factors $g \to \bar{g} = \psi^{-2} g$ which never vanish. This requires the following restrictions:

a) $|c| > 1$

b) $\psi(u) = \frac{1}{2}u^2 + c > 0$ for $c > 0$.

c) $\psi(u) = \cos h \; u + c > 0$ for $c > -1$
$\psi(u) = e^u + c > 0$ for $c \geq 0$.

This leads to global conformal transformations into (not necessarily onto) the following spaces:

a) $S^n(1) \to S^n(c^2 - 1)$
$\psi(u) = \sin u + c \quad , \; |c| > 1$

b) $E^n \to S^n(2c)$
$\psi(u) = \frac{u^2}{2} + c \quad , \; c > 0$

$$\text{c)}\quad H^n(-1) \to \begin{cases} S^n(1-c^2) & \text{for } |c| < 1 \\ E^n & \text{for } c = 1 \\ H^n(1-c^2) & \text{for } c > 1 \end{cases}$$

$$\psi(u) = \cos h\, u + c \quad , c > -1 \quad .$$

$$H^n(-1) \to E^n$$
$$\psi(u) = e^u + c \qquad , c \geq 0$$

In each case the curvature of the conformally transformed space is constant according to proposition 3 , and this curvature can be computed from the formula

$$\bar{K} = \psi^2 K + \frac{2}{n}\psi\, \Delta\psi - \|\nabla\psi\|^2 \quad .$$

Example 2 . <u>Einstein spaces which are not of constant sectional curvature</u>

a) Let (M_*, g_*) be a complete $(n-1)$-dimensional Einstein space $(n \geq 5)$ which is not of constant sectional curvature and whose scalar curvature is $\rho_* = -1$. Then $(\mathbb{R} \times M_*, g)$ with

$$ds^2 = du^2 + \cos h^2 u\, ds_*^2$$

is a complete Einstein space of scalar curvature $\rho = -1$ by lemma 13 (iii) .

This admits the globally defined function

$$\psi(u) = \sin h\, u + c$$

satisfying

$$\nabla^2\psi = \psi \cdot g \quad .$$

However, it is not a globally defined conformal factor.

Compare the example in [B-K] .

b) Let (M_*, g_*) be as above but with scalar curvature $\rho_* = 0$ (Ricci flat). Then $(\mathbb{R} \times M_*, g)$ with

$$ds^2 = du^2 + e^{2u}\, ds_*^2$$

is a complete Einstein space of scalar curvature $\rho = -1$, and the function

$$\psi(u) = e^u + c \quad (c \geq 0)$$

is globally defined and nowhere zero. Therefore it induces a globally defined conformal transformation

$$g \mapsto \bar{g} = \psi^{-2} g$$

where $(\mathbb{R} \times M_*, \bar{g})$ is a non-complete Einstein space with

$$d\bar{s}^2 = (e^u + c)^2 du^2 + \left(\frac{e^u}{e^u + c}\right)^2 ds_*^2 \quad .$$

The scalar curvature of $\bar{g}$ can be computed by lemma 1(iv) :

$$\bar{\rho} = \psi^2 \rho + \frac{2}{n}\, \psi\, \Delta\psi - \|\nabla\psi\|^2$$

$$= -(e^u + c)^2 + 2(e^u + c)e^u - e^{2u} = -c^2 \quad .$$

In the special case $c = 0$ we get

$$d\bar{s}^2 = e^{-2u}\, du^2 + ds_*^2$$

which is indeed the Ricci-flat Riemannian product $(0,\infty) \times M_*$. For examples of complete Ricci flat metrics compare [Be] ch. 15 .

Example 3 . Non-Einstein spaces admitting concircular transformations

a) The simplest example in dimension $n \geq 3$ is the metric g with

$$ds^2 = du^2 + e^{2u}\, ds_1^2$$

defined on $M = \mathbb{R} \times S^2$.

The function $\psi(u) = e^u$ satisfying

$$\nabla^2\psi = \psi \cdot g$$

induces a globally defined concircular transformation

$$(M,g) \to (M,\bar{g}) \quad , \quad \bar{g} = \psi^{-2}g \quad .$$

As in Example 2b) g is complete and $\bar{g}$ is not. g is not of constant scalar curvature because the radial curvature is -1 (see 11 (v)) and the sectional curvature in the plane σ orthogonal to $\frac{\partial}{\partial u}$ is by lemma 13 (i)

$$K_\sigma = e^{-2u}(1 - e^{2u}) = e^{-2u} - 1$$

which is non-constant.

This is a counterexample to the main result in [DH] .

b) Let (M_*, g_*) be an arbitrary complete $(n - 1)$-manifold, and let $\psi(u)$ be a smooth function $\psi : \mathbb{R} \to \mathbb{R}$ satisfying

(i) $\psi(u) > 0$ and $\psi'(u) > 0$ for all $u \in \mathbb{R}$,

(ii) $\int\limits_0^\infty \frac{du}{\psi(u)} = +\infty$

(iii) $\lim\limits_{n\to-\infty} \psi(u) = 0$

The metric

$$ds^2 = du^2 + (\psi'(u))\, ds_*^2$$

is a complete metric on $M = \mathbb{R} \times M_*$. By 12 ψ satisfies $\nabla^2\psi = \frac{\Delta\psi}{n} g$, thus ψ induces a globally defined concircular transformation $(M,g) \to (M,\bar{g})$, $\bar{g} = \psi^{-2}g$. $\bar{g}$ is again complete because the geodesic tangent to $\frac{\partial}{\partial u}$ has infinite length:

$$\int_0^{\infty} \left(\bar{g}\left(\frac{\partial}{\partial u}, \frac{\partial}{\partial u}\right)\right)^{1/2} du = \int_0^{\infty} \frac{du}{\psi(u)} = +\infty \quad ,$$

$$\int_{-\infty}^{0} \left(\bar{g}\left(\frac{\partial}{\partial u}, \frac{\partial}{\partial u}\right)\right)^{1/2} du = \int_{-\infty}^{0} \frac{du}{\psi(u)} = +\infty$$

where the last equality follows from $\lim_{u\to-\infty} \psi(u) = 0$.

This may serve as a counterexample to lemma 4.3 in [T 3] and proposition 6.2 in [T 4].

The choice of the level hypersurface M_* of ψ is absolutely arbitrary in this case, except that M_* should be complete if we want (M,g) and $(M,\bar{g})$ to be complete.

Note that this example cannot be modified such that both spaces have constant scalar curvature, see proposition 4 .

Example 4 . Special equations on surfaces

a) A complete 2-manifold admitting a nonconstant solution ψ of $\nabla^2\psi = K\cdot g$.

This example is due to R. Walter, and it is a counterexample to Satz II.1.13 in H. Wissner's thesis (Erlangen 1978).

We define $(\mathbb{R}^2, g)$ by

$$ds^2 = du^2 + \tanh^2 u \, dv^2$$

in polar coordinates (u,v) . This is a complete metric, and the function

$$\psi(u) = 2 \log(\cosh u)$$

satisfies

$$\nabla^2\psi = \psi''\cdot g \qquad \text{with} \quad \psi''(u) = \frac{2}{\cosh^2 u} \quad .$$

On the other hand the Gaussian curvature K is given by

$$K(u) = -\frac{\psi'''(u)}{\psi'(u)} = \frac{2}{\cosh^2 u} \quad .$$

This implies $\nabla^2\psi = K\cdot g$. The equation $\psi'' = K$ is essentially a Riccati equation $\psi'^2 + \psi'' = \text{constant}$.

b) A complete 2-manifold admitting a nonconstant solution of $\nabla^2\psi = K\cdot\psi\cdot g$. In the metric $ds^2 = du^2 + (\psi'(u))^2\, dv^2$ the equation $\nabla^2\psi = K\cdot\psi\cdot g$ reads as

$$\psi'' = -\frac{\psi'''}{\psi'}\cdot\psi$$

or equivalently $\psi\cdot\psi'' = \text{constant}$.

An explicit solution can be described for its inverse function $u(\psi)$:

$$u(\psi) = \pm\int_1^{\psi} \frac{d\xi}{\sqrt{2\log\xi}} \quad , \quad \psi \geq 1 \quad .$$

It follows that $\psi(0) = 1$, $\psi'(0) = 0$, $\psi''(0) = 1$ and that this is a complete metric $(\mathbb{R}^2, g)$ in polar coordinates (u,v) .

Note that the similar equation $\nabla^2\psi = -K\psi\cdot g$ or equivalently $\psi'' = -K\cdot\psi$ implies

$$\frac{\psi'}{\psi} = \frac{\psi'''}{\psi''} \quad \text{or} \quad \log\frac{\psi}{\psi''} = \text{constant} \quad .$$

It follows that K must be constant.

F. Solution of the global problem

In this section we discuss the following global problems:

I. (compare the introduction above)
Which complete Einstein spaces can be mapped conformely onto another (possibly non-complete) Einstein space?

II. Which complete Einstein spaces can be mapped conformally onto another complete Einstein space?

III. Which complete Riemannian manifolds admit a globally defined concircular mapping onto another manifold?

IV. Which complete Riemannian manifolds admit a globally defined solution of the equation $\nabla^2\psi = \frac{\Delta\psi}{n}\cdot g$?

All the problems are dealing with the equation $\nabla^2\psi = \frac{\Delta\psi}{n}\cdot g$. Consequently problem IV is the first one to be solved.

21. Theorem (compare [T 3] Lemma 2.2). Let (M^n,g) be a complete connected Riemannian manifold admitting a nonconstant solution ψ of $\nabla^2\psi = \frac{\Delta\psi}{n}\cdot g$. Then the number of critical points of ψ is $N \leq 2$, and M is conformally diffeomorphic to

(i) the sphere (S^n,g_1) if $N = 2$,
(ii) the euclidean space (E^n,g_o) or hyperbolic space (H^n,g_{-1}) if $N = 1$,
(iii) the Riemannian product $I \times M_*$ if $N = 0$, where (M_*,g_*) is a complete $(n-1)$-manifold and $I \subseteq \mathbb{R}$ is an open interval.

Vice versa each of these cases occurs (case (iii) with arbitrarily given (M_*,g_*)).

As part of the proof we show first the following lemma:

22. Lemma. Let (M^n,g) and ψ be as in theorem 21. If P denotes the set of critical points of ψ then $N := |P| \leq 2$, and $(M\setminus P,g)$ is isometric with a warped product metric

(1) $$ds^2 = du^2 + (\psi'(u))^2 ds_*^2$$

on $I \times M_*$ where (M_*,g_*) is a complete $(n-1)$-manifold and $I \subseteq \mathbb{R}$ is an open interval.

Proof of the lemma. By lemma 18 P is a set of isolated points. By lemma 12 for every fixed point $q \in M \setminus P$ there is an open neighborhood U such that (1) holds in U where ds_*^2 is the line element of the level hypersurface $M_* := \{x \in M \mid \psi(x) = \psi(q)\}$. (M_*,g_*) must be complete because every Cauchy sequence in M_* must converge in M . Consequently we may assume that $U = (\alpha,\beta)\times M_*$.

The trajectory through q is the unique geodesic with tangent $\frac{\partial}{\partial u}$. By completeness this is defined for every parameter u.

Now define α_o and β_o to be the infimum and supremum of α, β such that (1) holds for $(\alpha, \beta) \times M_*$.

If α_o (or β_o) is finite then there is a limit point q_o on this geodesic with $\psi(q_o) = \psi(\alpha_o)$ (or $\psi(\beta_o)$). Then q_o must be a critical point of ψ because otherwise lemma 12 would yield a contradiction. q_o is a minimum (a maximum) of ψ because by lemma 13 $\nabla^2\psi$ is definite at a critical point. There are the following possibilities:

I. $\alpha_o, \beta_o \in \mathbb{R}$, then $(\alpha_o, \beta_o) \times M_*$ with the warped product metric (1) is a compact manifold if we add the two critical points of level $\psi(\alpha_o)$ and $\psi(\beta_o)$. By connectness no other critical points can occur, hence $N = 2$.

II.a) $\alpha_o \in \mathbb{R}, \beta_o = +\infty$, then $(\alpha_o, \infty) \times M_*$ is complete if we add the minimum of level $\psi(\alpha_o)$, hence $N = 1$.

II.b) $\alpha_o = -\infty, \beta_o \in \mathbb{R}$, then the same holds for $(-\infty, \beta_o) \times M_*$.

III. $\alpha_o = -\infty, \beta_o = +\infty$, then $\mathbb{R} \times M_*$ is complete with the warped product metric (1), hence $M \cong \mathbb{R} \times M_*$, and $N = 0$.

<u>Proof of the theorem.</u> <u>1st case.</u> $N = 0$. Then ψ induces globally a product decomposition $M \cong \mathbb{R} \times M_*$, and the metric g satisfies $ds^2 = du^2 + (\psi'(u))^2 ds_*^2$ globally.

By a conformal change the metric transforms into

$$d\bar{s}^2 = \frac{1}{(\psi'(u))^2} du^2 + ds_*^2$$

which is a Riemannian product $I \times M_*$, I being an open interval in $\mathbb{R}$.

<u>2nd case.</u> $N = 1$. Then by lemma 22 the metric g satisfies $ds^2 = du^2 + (\psi'(u))^2 ds_*^2$ in $M \setminus \{p\}$ where p is the only critical point. By lemma 17 the levels near p must

be isometric with the standard sphere of certain radius, hence every level must be such a sphere and it follows that

$$ds^2 = du^2 + \frac{(\psi'(u))^2}{(\psi''(0))^2} ds_1^2 \quad .$$

If we regard this as an expression in polar coordinates around p we see that M is diffeomorphic to $\mathbb{R}^n$. By a conformal change the metric transforms into the euclidean or hyperbolic metric depending on the growth of $\psi'(u)$ when u tends to infinity.

<u>3rd case.</u> $N = 2$. Then M is compact, and by the same argument the metric satisfies

$$ds^2 = du^2 + \frac{(\psi'(u))^2}{(\psi''(0))^2} ds_1^2$$

in $M \setminus \{p,q\}$ where p,q are the critical points. It follows that M is diffeomorphic to the sphere. Locally the metric can be conformally transformed into a metric of constant curvature. Hence (M,g) is a compact conformally flat manifold. By a theorem of N.H. Kuiper [Kui 1] (M,g) is conformally equivalent to (S^n,g_1).

To see that all these cases really occur, compare the examples in section E .

This completes the proof of theorem 21 .

This solves problem IV . In order to get examples for case (iii) one may prescribe arbitrarily a complete $(n-1)$-manifold (M_*,g_*) and a smooth (or C^2) function $\psi : \mathbb{R} \to \mathbb{R}$ with the only condition that ψ' is never zero. The solution of problem III is the same by proposition 8 with the only extra condition that ψ should not be zero. See our examples 3a) and b) in section E .

23. <u>Corollary.</u> <u>Let</u> (M,g) <u>be a compact Riemannian manifold admitting a nonconstant solution</u> ψ <u>of</u> $\nabla^2\psi = \frac{\Delta\psi}{n}\cdot g$. <u>Then</u> (M,g) <u>is conformally diffeomorphic to</u> (S^n,g_1) .

This follows from 21 because the compactness forces ψ to have at least two critical points.

The situation is quite different if we put additional restrictions on the metric g. For compact manifolds we have the following:

24. Theorem. Let (M,g) be a compact Riemannian manifold (without boundary) of constant scalar curvature. Assume that it admits a non-constant solution ψ of $\nabla^2\psi = \frac{\Delta\psi}{n}\cdot g$. Then (M,g) is isometric with the standard sphere of certain radius.

Proof. From 23 we know that (M,g) is conformally diffeomorphic to the sphere. From lemma 22 we get in $M \setminus \{p,q\}$ the expression $ds^2 = du^2 + \frac{(\psi'(u))^2}{(\psi''(0))^2} ds_1^2$ where ds_1^2 is the line element of the standard unit sphere.

Then lemma 13 implies that the scalar curvatures ρ = constant of M and $\rho_* = (\psi''(0))^2$ of the level hypersurface M_* satisfy the equation

$$\psi'^2\cdot\rho = \frac{n-2}{n}\rho_* - \frac{n-2}{n}\psi''^2 - \frac{2}{n}\psi'\psi''' .$$

Let us regard this as a differential equation for the unknown function $\psi' = \psi'(u)$ with the initial conditions $\psi'(0) = 0,\ \psi''(0) = \sqrt{\rho_*}$.

If $\rho > 0$ then one solution is easily seen:

$$\psi'(u) = \sqrt{\rho_*/\rho}\cdot\sin(\sqrt{\rho}\cdot u) .$$

On the other hand the equation is nonlinear and the solution is possibly not unique in the singularity $\psi' = 0$.

In the following we will integrate the equation by elementary operations. Let us write $y := \psi'$, then the equation reads as

$$yy'' + \frac{n-2}{2}y'^2 + \frac{n}{2}\rho\cdot y^2 - \frac{n-2}{2}\rho_* = 0$$

(compare [Kam] 6.224).

We multiply it by $2y^{n-3}y'$ and get

$$(y^{n-2}y'^2)' + (\rho y^n - \rho_* y^{n-2})' = 0$$

or equivalently

$$y^{n-2}y'^2 + \rho y^n - \rho_* y^{n-2} = \text{constant.}$$

The initial condition implies that this constant must vanish:

$$y^{n-2}(y'^2 + \rho y^2 - \rho_*) = 0 \quad .$$

This implies that a solution with our initial conditions must satisfy

$$y'^2 + \rho y^2 - \rho_* = 0$$

for all sufficiently small parameters $u > 0$. Differentiating again we get

$$2y'(y'' + \rho y) = 0$$

which implies

$$\psi'(u) = y(u) = \begin{cases} c \cdot \sin(\sqrt{\rho}\, u) & \text{for } \rho > 0 \\ c \cdot u & \text{for } \rho = 0 \\ c \cdot \sin h(\sqrt{-\rho}\, u) & \text{for } \rho < 0 \end{cases} ,$$

where c is a constant.

Then 13(iv) implies that (M,g) is a space of constant sectional curvature ρ . We know already that M is diffeomorphic with the sphere. This shows that $\rho > 0$ and that (M,g) is isometric with the standard sphere of radius $1/\sqrt{\rho}$.

Concircular transformations between spaces of constant scalar curvature and conformal transformations between Einstein spaces are characterized by the equation

$$\nabla^2\psi = (-\rho\psi + B)\cdot g \qquad , \rho, B \in \mathbb{R}$$

(compare lemma 2 , Proposition 4 and 8).

This suggests to study this special type of the equation on (M,g) where ρ and B are constants (not assumed to be related with the metric g).

25. <u>Theorem</u> (Obata [Ob 2], Tashiro [T 3]). <u>Let</u> c, B <u>be constants and</u> (M,g) <u>a complete Riemannian manifold admitting a nonconstant solution of</u> $\nabla^2\psi = (-c^2\psi + B)\cdot g$. <u>Then</u> (M,g) <u>is isometric with the standard sphere of radius</u> $1/c$.

<u>Proof.</u> Let P be the set of critical points of ψ . From lemma 22 we have the global expression in $M \setminus P$

$$ds^2 = du^2 + (\psi'(u))^2 ds_*^2$$

where ψ is a function depending on the arclength parameter on the geodesic trajectory and ds_*^2 is the metric of a complete level hypersurface. The equation $\nabla^2\psi = (-c^2\psi + B)\cdot g$ reads as $\psi'' = -c^2\psi + B$.

It follows that

$$\psi(u) = a\cos(cu) + b\sin(c\,u) + \frac{B}{c^2}$$

where a,b are certain constants.

In particular ψ must have at least two critical points (minimum and maximum) because ψ is bounded. Let $u = 0$ be one of these critical points. Then lemma 18 implies that the levels are standard spheres and that

$$ds^2 = du^2 + \frac{(\psi'(u))^2}{(\psi''(0))^2}\, ds_1^2 \ .$$

From $\psi'(0) = 0$ we get $b = 0$ and $ds^2 = du^2 + c^2\sin^2(cu)\cdot ds_1^2$ which is the standard metric on the sphere of radius $1/c$.

26. Proposition (Tashiro [T 3]). Assume that a complete manifold admits a nonconstant solution of $\nabla^2\psi = B\cdot g$, $0 \neq B \in \mathbb{R}$. Then it is isometric with the euclidean space.

Proof. As in the proof of 25 we have in $M \setminus P$

$$ds^2 = du^2 + (\psi'(u))^2 ds_*^2$$

and $$\psi'' = B \quad .$$

It follows that $\psi(u) = \frac{B}{2} u^2 + bu + c$ with constants b,c . Because of $B \neq 0$ ψ must have a critical point and by lemma 18 ds_*^2 is the metric of a standard sphere. Let $u = 0$ be the critical point, then $b = 0$, $ds_*^2 = \frac{1}{B^2} ds_1^*$ and therefore

$$ds^2 = du^2 + u^2 ds_1^2$$

which is the metric of the euclidean space in polar coordinates.

Remark. A similar characterization theorem is not possible using the equation $\nabla^2\psi = 0$ or $\nabla^2\psi = (c^2\cdot\psi + B)\cdot g$ because in these cases solutions without critical points are possible. In these cases the function ψ can be determined but not the metric of the level hypersurfaces of ψ .

Now we want to give the solution of problem I and II .

27. Main Theorem. (i) Let (M,g) be a complete Einstein space. Then there exists a non-homothetic conformal diffeomorphism $(M,g) \to (\bar{M},\bar{g})$ onto another (possibly non-complete) Einstein space if and only if (M,g) is isometric with one of the examples 1a)b)c) or 2b) in section E , i.e. if M is either a sphere, euclid-

<u>ean or hyperbolic space or the product</u> $M = \mathbb{R} \times M_*$ <u>with a complete Ricci flat space</u> (M_*, g_*) , <u>endowed with the warped product metric</u>

$$ds^2 = du^2 + e^{2u} ds_*^2 \qquad \text{(up to scaling)}$$

<u>which has negative scalar curvature.</u>

(ii) <u>If in addition</u> $(\bar{M}, \bar{g})$ <u>is complete then each of the spaces is isometric with a standard sphere of certain radius.</u>

<u>Remark.</u> Concerning part (ii) it has been shown in 1950 by N.H. Kuiper [Kui 2] that the scalar curvature must be positive. K. Yano and T. Nagano [Y-N] have shown (ii) for the case of a 1-parameter group of conformal transformations of M onto itself, and (ii) has been shown by R. Kulkarni [Kul] for one conformal transformation of M onto itself. T. Nagano [Na] stated (ii) under the weaker assumption of parallel Ricci tensor.

The history about part (i) is strange enough. Y. Tashiro [T 3] had all arguments to prove (i) but he did not formulate it. Later in [T 4] he gave an incorrect version of (i) . We have not been able to find a correct version of (i) in the literature. Theorem G in [Kan] is quite close but it is not really satisfying in some point. Compare our discussion of the "standard wrong theorems" below.

<u>Proof.</u> (i) We compare the metrics g and $\frac{1}{\psi^2} \cdot g = \tilde{g} := f^* \bar{g}$ on M . By lemma 2 $\psi : M \to \mathbb{R}$ satisfies the equation $\nabla^2 \psi = (-\rho\psi + B) \cdot g$. By lemma 22 there are at most two critical points of ψ , and the metric g satisfies

$$ds^2 = du^2 + (\psi'(u))^2 ds_*^2$$

except in the critical points. Here ds_*^2 is a complete metric on an $(n - 1)$-manifold M_* , and (M_*, g_*) is isometric with a standard sphere if there is a critical point of ψ (lemma 18).

If ψ has a critical point p then (M,g) is a space of constant sectional curvature near p by 19 . By the global warped product metric it follows that (M,g) is globally a space of constant sectional curvature. Then 21 says that it is in fact one of the standard spaces S^n, E^n, H^n . Compare the examples 1a)b)c) in section E .

It remains to discuss the case where ψ has no critical points. The differential equation for $\psi : \mathbb{R} \to \mathbb{R}$ is

$$\psi'' = -\rho\psi + B$$

(compare 14). We know in addition that $\psi(u) \neq 0$ for all $u \in \mathbb{R}$ and of course $\psi'(u) \neq 0$ for all $u \in \mathbb{R}$. This is possible if and only if $\rho < 0$ and $\psi(u) = e^{\pm\sqrt{-\rho}u} + c$ with $c \geq 0$. Then the metric can be globally expressed by

$$ds^2 = du^2 - \rho e^{2\sqrt{-\rho}u} ds_*^2 .$$

Lemma 13 says that (M_*, g_*) must be an Einstein space with scalar curvature

$$\rho_* = \psi''^2 + \frac{2}{n-2}\psi'\psi''' + \frac{n}{n-2}\cdot\rho\cdot\psi'^2$$

$$= \rho^2 e^{2\sqrt{-\rho}u}\left(1 + \frac{2}{n-2} - \frac{n}{n-2}\right) = 0 .$$

This is exactly our example 2 b) in section E . In the special case $(M_*, g_*) = (E^{n-1}, g_o)$ we get the hyperbolic space of curvature ρ .

(ii) From (i) we know all candidates for (M,g) . We have to examine which metrics $\tilde{g} = \frac{1}{\psi^2}\cdot g$ are again complete. By scaling we may reduce the problem to the cases where $\rho = +1, 0, -1$.

Then the solutions of $\psi'' = -\rho\psi + B$ are simply

$$\psi(u) = \begin{cases} a \sin u + b \cos u + c & , \rho > 0 \\ a u^2 + bu + c & , \rho = 0 \\ a \sinh u + b \cosh u + c & \rho < 0 \end{cases}$$

where a,b,c are certain constants.

We consider a geodesic tangent to $\frac{\partial}{\partial u}$. By assumption it has infinite length in g. We compute

$$\tilde{g}\left(\frac{\partial}{\partial u},\frac{\partial}{\partial u}\right) = \frac{1}{(\psi(u))^2} = \begin{cases} 1/(au^2 + bu + c)^2 & , \rho = 0 \\ 1/(a\sinh u + b\cosh u + c)^2 , & \rho < 0 \end{cases}$$

In any of these cases either $\int_{-\infty}^{0}\left(\tilde{g}\left(\frac{\partial}{\partial u},\frac{\partial}{\partial u}\right)\right)^{1/2}du$ or $\int_{0}^{\infty}\left(\tilde{g}\left(\frac{\partial}{\partial u},\frac{\partial}{\partial u}\right)\right)^{1/2}du$ is finite which contradicts the completeness of $(M,\tilde{g})$.

It remains the case $\rho > 0$. Then (M,g) is isometric with a standard sphere and consequently $(\bar{M},\bar{g})$ must be isometric with a standard sphere (compare theorem 24 and 25).

This completes the proof of theorem 27. Note that in [Be] chapter 9. J a description of all complete Einstein spaces is sketched which have a warped product metric with 1-dimensional basis. However, theorem 27 does not follow directly from [Be] 9.113. The restrictions for the conformal factor ψ have to be analyzed in addition.

Appendix. Some standard wrong theorems

Standard wrong theorem I. Assume that a complete Riemannian manifolds (M,g) admits a nonconstant solution ψ of $\nabla^2\psi = \frac{\Delta\psi}{n} g$. Then M is diffeomorphic to the sphere.

References are:
1. [Y 2] thm. I ,
2. [Y 3] chapter 1 thm. 6.3
3. [Y-Ob] thm. E
4. [H-M] thm. A
5. [H-S] thm. C
6. [A-H] thm. 1.5 .

Slightly different versions (also wrong) are

7. [Si] lemma 4
8. [DH] main theorem.

Curiously enough the euclidean space (Ex. 1a in section E above) provides a counterexample to 1.-7. . For a counterexample to 8. see our Example 3a) . More sophisticated counterexamples may be constructed following the pattern of our Example 2a) and 3b) . These may be chosen such that the metric is not locally conformally flat. Therefore the conclusion "M is conformal to the sphere" in most of these theorems is not true, neither globally nor locally.

Standard wrong theorem II. Let (M,g) be a complete Einstein space admitting a conformal (or concircular) transformation into another Einstein space. Then (M,g) is isometric to either S^n, E^n or H^n (standard spaces of constant sectional curvature).

Note that the assumptions "conformal" and "concircular" are equivalent by Corollary 10 .

References are:
1. [I-T] thm. 3 (even more general)
2. [T 2] thm. B and thm. 2
3. [T 4] proposition 5.1 , 5.2
4. [Yau] corollary 4.1

A counterexample is given by our Example 2b) (compare theorem 27.).

REFERENCES

[A-H] K. AMUR and V.S. HEDGE, Conformality of Riemannian manifolds to spheres, J. Diff. Geom. 9 (1974), 571-576.

[Ba] C. BARBANCE, Transformations conformes d'une variété riemannienne compact, C.R. Acad. Sc. Paris 260 (1965), 1547-1549.

[B-K] C. BARBANCE and Y. KERBRAT, Sur les transformations conformes des variétés d'Einstein, C.R. Acad. Sc. Paris 286 (1978), 391-394.

[Be] A.L. BESSE, Einstein manifolds, Springer, Berlin-Heidelberg-New York (1987), Ergebnisse der Mathematik und ihrer Grenzgebiete, 3. Folge, Band 10)

[Br 1] H.W. BRINKMANN, Riemann spaces conformal to Einstein spaces, Math. Ann. 91 (1924) , 269-278.

[Br 2] H.W. BRINKMANN, Einstein spaces which are mapped conformally on each other, Math. Ann. 94 (1925), 119-145.

[DH] DǍNG-VŨ-HUYẾN, Variétés riemanniennes admettant une fonction u telle que $\nabla^2 u + f\, u\, g = 0$, Acta Math. Vietnam. 3 (1978), 17-21.

[Fe] J. FERRAND, Concircular transformations of Riemannian manifolds, Ann. Acad. Sci. Fenn. Ser. A.I. 10(1985),163-171.

[Fi] A. FIALKOW, Conformal geodesics, Transactions AMS 45 (1939), 443-473

[H-M] C.-C. HSIUNG and L.R. MUGRIDGE, Riemannian manifolds admitting certain conformal changes of metric, Coll. Math. 26 (1972), 135-143.

[H-S] C.-C. HSIUNG and L.W. STERN, Conformality and isometry of Riemannian manifolds to spheres, Transactions AMS 163 (1972), 65-73.

[I-T] S. ISHIHARA and Y. TASHIRO, On Riemannian manifolds admitting a concircular transformation, Math. J. Okayama Univ. 9 (1959), 19-47.

[Kam] E. KAMKE, Differentialgleichungen, Lösungsmethoden und Lösungen I , Teubner, Stuttgart (1977), 9. Auflage

[Kan] M. KANAI, On a differential equation characterizing a Riemannian structure of a manifold, Tokyo J. Math. 6 (1983), 143-151.

[Kui 1] N.H. KUIPER, On conformally-flat spaces in the large, Ann. Math. 50 (1949), 916-924.

[Kui 2] N.H. KUIPER, Einstein spaces and connections I,II, Indagationes math. 12 (1950), 505-521.

[Kul] R.S. KULKARNI, Curvature structures and conformal transformations, J. Diff. Geom. 4 (1969), 425-451.

[La 1] J. LAFONTAINE, Sur la géométrie d'une généralisation de l'équation différentielle d'Obata, J. Math. Pures et Appliquées 62 (1983), 63-72.

[La 2] J. LAFONTAINE, Conformal geometry from the Riemannian view-point, in this volume

[Na] T. NAGANO, The conformal transformation on a space with parallel Ricci tensor, J. Math. Soc. Japan 11 (1959), 10-14.

[Ob 1] M. OBATA, Conformal transformations of compact Riemannian manifolds, Illinois J. Math. 6 (1962), 291-295.

[Ob 2] M. OBATA, Certain conditions for a Riemannian manifold to be isometric with a sphere, J. Math. Soc. Japan 14 (1962), 333-340.

[Ob 3] M. OBATA, Conformal transformations of Riemannian manifolds, J. Diff. Geom. 4 (1970), 311-333.

[R] B. RUH, Krümmungstreue Diffeomorphismen Riemannscher und pseudo-Riemannscher Mannigfaltigkeiten, Math. Z. 189 (1985), 371-391.

[Si] U. SIMON, Einstein spaces isometrically diffeomorphic to a sphere, manuscripta math. 22 (1977), 1-5.

[Ta] S. TACHIBANA, On concircular geometry and Riemann spaces with constant scalar curvatures, Tôhoku Math. J. 3 (1951), 149-158.

[T 1] Y. TASHIRO, On projective transformations of Riemannian manifolds, J. Math. Soc. Japan 11 (1959), 196-204.

[T 2] Y. TASHIRO, Remarks on a theorem concerning conformal transformations, Proc. Japan Acad. 35 (1959), 421-422.

[T 3] Y. TASHIRO, Complete Riemannian manifolds and some vector fields, Transact. AMS 117 (1965), 251-275.

[T 4] Y. TASHIRO, Conformal transformations in complete Riemannian manifolds, Publ. Study Group of Geometry Vol. 3 (1967)

[T-M] Y. TASHIRO and K. MIYASHITA, On conformal diffeomorphisms of complete Riemannian manifolds with parallel Ricci tensor, J. Math. Soc. Japan 23 (1971), 1-10.

[Ve 1] P. VENZI, The metric $ds^2 = F(u)du^2 + G(u)d\sigma^2$ and an application to the concircular mappings, Utilitas math. 22 (1982), 221-233.

[Ve 2] P. VENZI, Über konforme und geodätische Abbildungen, Result. Math. 5 (1982), 184-198.

[Vo 1] W.O. VOGEL, Kreistreue Transformationen in Riemannschen Räumen, Arch. Math. 21 (1970), 641-645.

[Vo 2] W.O. VOGEL, Einige Kennzeichnungen der homothetischen Abbildungen eines Riemannschen Raumes unter den kreistreuen Abbildungen, manuscripta math. 9 (1973), 211-228.

[Y 1] K. YANO, Concircular geometry I-V, Proc. Imp. Acad. Japan 16 (1940), 195-200, 354-360, 442-448, 505-511, ibid. 18 (1942), 446-451.

[Y 2] K. YANO, On Riemannian manifolds admitting an infinitesimal conformal transformation, Math. Z. 113 (1970), 205-214.

[Y 3] K. YANO, Integral formulas in Riemannian geometry, M. Dekker, New York (1970).

[Y-N] K. YANO and T. NAGANO, Einstein spaces admitting a one-parameter group of conformal transformations, Ann. Math. 69 (1959), 451-460

[Y-Ob] K. YANO and M. OBATA, Conformal changes of Riemannian metrics, J. Diff. Geom. 4 (1970), 53-72.

[Yau] S.T. YAU, Remarks on conformal transformations, J. Diff. Geom. 8 (1973), 369-381.

Topics in the Theory of Quasiregular Mappings

Seppo Rickmann

Contents

1. Introduction

The theory of quasiregular maps has turned out to form the right generalization of the geometric part of the theory of one complex variable analytic functions to real n-dimensional space. These maps can be described as quasiconformal maps witout the homeomorphism requirement and, consequently, they have in general branching. The most interesting geometric features of the theory of quasiregular maps are in general of global character. While many relatively strong and precise results of this nature exist, the connections to differential geometry for example are not well understood and there is much left for further research.

The purpose of these notes is on one hand to give some of the basic working tools in the theory of quasiregular maps and, on the other hand, to give proofs or at least outlines of proofs of some of the main theorems. Most of the geometric results in the theory are proved by certain inequalities for moduli of path families. This method is presented in Section 5.

In 1967 Zorič [Zo] proved that a locally homeomorphic quasiregular map of R^n into itself is always a homeomorphism if $n \geqq 3$. In Section 6 we prove a slightly more general result which gives Zorič's theorem as a corollary. Zorič also posed the question of the validity of a Picard's theorem on omitted values and this problem was for a long time one of the main open questions in the theory of quasiregular maps. This problem is now rather completely settled for dimension three and partly also for higher dimensions. We shall give a rather complete proof of a Picard's theorem in Section 7 for all dimensions. We shall there also discuss the sharpness of the result.

An interesting question about which rather little is known, apart from the Picard type results mentioned above, is the general problem of the existence of a nonconstant quasiregular map $f: M \longrightarrow N$ for given connected oriented Riemannian n-manifolds M and N. Some scattered results of this mapping problem are contained in Section 8 of these notes.

2. Definitions and history

The definition of a quasiregular map for the Euclidean n-space $R^n, n \geqq 2$, is the following. Let G be a domain in R^n and let $f: G \longrightarrow R^n$ be continuous. f is called quasiregular (qr) if (1) f belongs to the local Sobolev space $W^1_{n,loc}(G)$ i.e. f has first order weak partial derivatives which are locally in $L^n(G)$, and (2) there exists K, $1 \leqq K < \infty$, such that

$$(2.1) \qquad |f'(x)|^n \leqq K J_f(x) \quad \text{a.e.}$$

By continuity and (1) f has ordinary partial derivatives $D_i f(x)$ a.e. The linear map $f'(x)$ is defined by $f'(x)e_i = D_i f(x)$ (e_i is the standard ith basis vector in R^n), $|f'(x)|$ is the operator norm of $f'(x)$, and $J_f(x)$ is the Jacobian determinant. The smallest $K \geqq 1$ in (2.1) is called the outer dilatation $K_0(f)$. If f is quasiregular, also

$$(2.2) \qquad J_f(x) \leqq K' \inf_{|h|=1} |f'(x)h|^n \quad \text{a.e.}$$

for some K', $1 \leqq K' < \infty$. The smallest K' in (2.2) is the inner dilation $K_I(f)$ and $K(f) = \max(K_0(f), K_I(f))$ is called the (maximal) dilatation of f. If $K(f) \leqq K$, f is called K-quasiregular.

The definition of quasiregularity extends in a

straightforward manner to the case $f:M \to N$ where M and N are connected oriented Riemannian n-manifolds since both conditions (1) and (2) make sense. All manifolds are in these notes assumed to be connected and oriented. A qr homeomorphism is called a quasiconformal (qc) map. If $N = \overline{R}^n = R^n \cup \{\infty\}$, $M \subset \overline{R}^n$, and if $\overline{R}^n$ is equipped with the spherical metric, a qr map $f:M \longrightarrow N$ is also called quasimeromorphic. Throughout this article the letter n is designated for the dimension in connections with qr maps.

In the Euclidean case we have the following situation. For $n = 2$ the 1 - qr mappings are exactly the analytic functions, and a qr map f can always be written as a composition $f = g \circ h$ where h is qc and g analytic. For $n \geq 3$ the 1 - qr maps are very rigid. According to the generalized Liouville theorem they are restrictions of Möbius transformations. In this general form it is due to Gehring [G2] and Reshetnyak [Re2], see also [BI1]. In order to obtain an interesting theory for $n \geq 3$, we need the dilatation to be > 1.

Interesting qr maps for $n \geq 3$ have in general branching because of Zorič's theorem (cf. Introduction). Partly for this reason it is important not to restrict to too smooth maps in our definition of quasiregularity. For example qr maps in C^3 are locally homeomorphic for $n \geq 3$. The local Sobolev class $W^1_{n,loc}$ is the right class if we want compactness of families of K - qr maps.

The history of qr maps is as follows. Smooth qr

mappings for $n = 2$ were considered by Grötzsch in 1928. He proved for example the Picard's theorem for such maps. Lavrentjev studied smooth locally homeomorphic qr maps for $n \geq 3$ in 1938. He already stated Zorič's theorem without proof.

It was Reshetnyak who gave the definition as presented here and started in 1966 a systematic study of qr maps (by using the term "maps with bounded distortion") in a series of articles. His results are contained in the book [Re5]. One of Reshetnyak's main results is that a nonconstant qr map f is discrete (i.e. $f^{-1}(g)$ consists of isolated points) and open. A more geometric view of the subject was taken by the group Martio, Rickman, and Väisälä in the articles [MRV 1,2,3].

Reshetnyak's main method is nonlinear potential theory, more precisely, extremals of certain variational integrals serve as counterparts for the harmonic functions in the plane. This theory has been studied further by Granlund, Lindqvist, and Martio (see for example [GLM]), and also by Bojarski and Iwaniec [BI2].

Other names that should be mentioned in connection with qr maps are U. Srebro, E.A. Poleckii, J. Ferrand, J. Sarvas, M. Vuorinen, M. Gromov, and P. Pansu. Historical remarks are included also in subsequent sections, in particular, the history of the problem of Picard's theorem is included in Section 7.

3. Examples

3.1. Finite-to-one branched covers. One of the simplest non-trivial examples is the winding map $f: R^3 \longrightarrow R^3$, $f(r,\varphi,x_3) = (r,k\varphi,x_3)$ in cylinder coordinates where $k > 1$ is an integer. This map when extended to a map $\overline{R}^3 \longrightarrow \overline{R}^3$ belongs to a whole group of examples of qr maps, namely branched covers between compact n-manifolds. The winding map f has the property that the dilatation grows to infinity together with k.

Another example when the dilatation remains under a universal bound, but still the local index at some point is arbitrarily high, is constructed as follows. The local index $i(x,f)$ at x of a discrete open mapping f is defined as the number $\sup_y \operatorname{card} U \cap f^{-1}(y)$ where U is a sufficiently small neighborhood of x (this number is independent of the chosen U). Let now the boundary A of the cube $\{x \in R^3 : |x_j| \leq 1 \quad i = 1,2,3\}$ be triangulated as follows. We divide A first into any number of congruent squares and then divide each square into four (open) congruent triangles letting the vertices be those of the square plus the center of the square. Pick one triangle T in this triangulation. We can map the cone $C = \{\lambda T : \lambda > 0\}$ by a K-qc map f_0 onto the half space $H_+ = \{x \in R^2 | x_3 > 0\}$ where K does not depend on the mesh of the triangulation. By straightforward modified repeated reflection through the sides of the cones corresponding to the triangles and through ∂H_+ we extend f_0 to a K_1-qr map $f: R^3 \longrightarrow R^3$ where K_1 depends only on K. Here the local index $i(0,f)$ can be made arbitrarily large.

3.2. Zorič's map. In [Zo] Zorič gave the following example of a qr map of R^3 into $R^3 \setminus \{0\}$ which can be regarded as the counterpart of the exponential function in the plane. Let C_0 be the cylinder $\{x \in R^3 : 0 < x_1, x_2 < 1\}$. We can again map C_0 by a K-qc map f_0 onto the half space H_+ such that the edges of C_0 correspond to rays emanating from the origin. We can extend f_0 by repeated reflections through faces of C_0 and ∂H_+ and obtain a qr mapping $f: R^3 \longrightarrow R^3 \setminus \{0\}$. This example as well as those in 3.1 can be generalized to higher dimensions in a straightforward manner.

3.3. Automorphic qr maps. Zorič's map f is an example of a qr map that is automorphic with respect to a Möbius group Γ, that is, $f \circ g = f$ for all g in Γ. General existence results have been proved for discrete Möbius groups acting on the unit ball B^n. For example in [MS] it is shown that if the group has finite volume there exists always an automorphic qr map with respect to the group.

4. Basic properties

In this section we shall present some basic facts of qr maps which are relevant for later discussions, mostly without proofs.

4.1. Reshetnyak's main theorem [Re1],[Re3]. *Every nonconstant qr map is discrete and open.*

This shows that qr maps also for $n \geqq 3$ share the topological characterization with the analytic functions.

The essential step in Reshetnyak's proof is to show that point inverses $f^{-1}(y)$ for a map $f:G \longrightarrow R^n$ have zero n-capacity (see Section 5 for this concept). This he achieved by showing that $\log|f|$ is an extremal for a variational integral. Let it be remarked here that the corresponding Euler equation is a quasilinear second order PDE, and in this sense there is a strong interaction between qr and PDE theory. Recently an improved proof arrangement of 4.1 was given in [BI2].

If f is a discrete open map between topological n-manifolds, it is well-known by results of Černavskii in 1964 and Väisälä in 1966 that

$$(4.2) \qquad \dim B_f = \dim fB_f \leqq n - 2$$

where B_f is the branch set, i.e. the set where f fails to be a local homeomorphism and dim means topological dimension.

On the other hand, the (n-2)-dimensional Hausdorff measure of fB_f satisfies

$$(4.3) \qquad H^{n-2}(fB_f) > 0 \quad \text{if} \quad B_f \neq \phi$$

for a discrete open map f [MRV3]. For the branch set itself this is known only for $n = 2,3$.

Next we consider instances where $B_f = \phi$ for qr maps. First we take up the implication of smoothness mentioned in Section 2.

4.4. Proposition. *Let* $n \geq 3$ *and let* $f : G \longrightarrow R^n$ *be qr. Suppose*

$$f \in C^3 \quad \text{if} \quad n = 3 \, ,$$
$$f \in C^2 \quad \text{if} \quad n \geq 4 \, .$$

Then $B_f = \phi$ *or* f *is constant.*

Proof. If $f \in C^1$ and $x \in B_f$, then $f'(x) = 0$. By [F,3.4.3] we have therefore $H^{n/k}(fB_f) = 0$ whenever $f \in C^k$, $k \geq 1$. If f is not constant, 4.1 and (4.3) give the result.

It is an open question whether the assumption $f \in C^1$ is enough in 4.4. The following result reflects in some sense a stability feature in Liouville's theorem.

4.5. Theorem [MRV, 4.6] *For every* $n \geq 3$ *there exists* $K_n > 1$ *such that every nonconstant* K_n*-qr map is a local homeomorphism.*

The results 4.4 and 4.5 are not true in the plane. The analytic functions serve as counter examples. In connection with 4.5 there is also an open problem, namely no explicit bound > 1 is known for K_n.

4.6 Theorem. *Let* $f : M \longrightarrow N$ *be nonconstant qr. Then*

(1) [Re1] f *is differentiable a.e.*

(2) [Re1] f *satisfies the condition* (N), *i.e.* $m(fA) = 0$ *whenever* $m(A) = 0$. *Here* m *is the Lebesgue measure.*

(3) [MRV1,2.27] $m(fB_f) = 0$.

(4) *If* $h : N \longrightarrow [0,\infty]$ *and* $E \subset M$ *are measureable,*

$$\int_E (h\circ f) J_f dm = \int_N h(y) N(y,f,E)\,dy.$$

Here $N(y,f,E) = \operatorname{card} f^{-1}(y) \cap E$.

(5) [MRV1,8.2] $J_f(x) > 0$ _a.e._

(6) [MRV1,8.3] $m(B_f) = 0$.

We shall give the proofs of (5) and (6) in the end of Section 5 as an application of the method of moduli of path families. For a proof of (4) in this situation, see [Pe2].

Let $f: M \longrightarrow N$ be a nonconstant qr map. Then f is also sense-preserving in the sense that for $D \subset\subset M$ and $y \in fD \setminus f\partial D$ the _topological degree_

$$\mu(y,f,D) = \sum_{x\in D\cap f^{-1}(y)} i(x,f) > 0,$$

see [MRV1]. In particular, $i(x,f) \geq 1$ for all $x \in M$. According to properties of discrete open maps, for each $x \in M$ there exists a neighborhood W of $f(x)$ such that $f|(U\setminus f^{-1}f(B_f\cap U))$ is a $i(x,t)$-to-one covering of $W\setminus f(B_f\cap U)$. Here U is the x-component of $f^{-1}W$.

5. Inequalities for moduli of path families

In this section we shall present the most effective tool for proofs in the qr theory for $n \geq 3$. The method of moduli of path families, or extremal length, has for a long time been used in conformal and qc theory. It was introduced by Beurling in 1946 and published in 1950 by Ahlfors and Beurling (for $n=2$). Later important contributions applicable to qc maps are due to Fuglede, Gehring, and Väisälä. A good reference is the notes [V1].

For qr maps there are two important inequalities, one is due to Poleckii and the other to Väisälä. We shall outline the proof of these, but omit the most technical part, the proof of the so called Poleckii's Lemma. In the end we shall give some applications.

Let now M be a Riemannian n-manifold and let Γ be a family of nonconstant paths in M. The (n-)modulus of Γ is defined as

$$(5.1) \qquad M(\Gamma) = \inf_{\rho} \int_M \rho^n dm$$

where the infimum is taken over all Borel functions $\rho : M \longrightarrow [0,\infty]$ such that the line integral satisfies

$$\int_{\gamma} \rho ds \geq 1$$

for all locally rectifiable $\gamma \in \Gamma$. The number $1/M(\Gamma)$ is called the extremal lenght of Γ. Hence $M(\Gamma)$ has the meaning of the volume of an "extremal" conformal metric such that all paths have length at least one. The locus (=image) of a path γ is denoted by $|\gamma|$ and the locus of Γ, which is the union of all $|\gamma|$, $\gamma \in \Gamma$, is denoted by $|\Gamma|$. The family of admissible functions ρ in (5.1) is denoted by $F(\Gamma)$.

5.2 Examples. (a) Let $A \subset R^{n-1} = R^{n-1} \times \{0\} \subset R^n$ be a Borel set, let $h > 0$, and let Γ be the family of line segments

$\gamma_y : [0,h] \longrightarrow R^n$, $\gamma_y(t) = y + te_n$, $y \in A$. Then

$$M(\Gamma) = \frac{m_{n-1}(A)}{h^{n-1}}$$

and the extremal ρ in (5.1) is $1/h$ in $B = A \times [0,h]$ and zero outside. For the larger family of paths joining A and $A + he_n$ in B one obtains the same modulus. For details, see [V1,7.2] .

(b) Let $0 < s < t$, let E be a Borel set in the unit sphere S^{n-1} and let Γ be the family of line segments $\gamma_y : [s,t] \longrightarrow R^n$, $\gamma_y(u) = uy$, $y \in E$. Then (see [V1,7.5])

$$M(\Gamma) = \frac{m_{n-1}(E)}{(\ln \frac{t}{s})^{n-1}}$$

and the extremal ρ in $|\Gamma|$ is now

$$\rho(x) = \frac{1}{(\ln \frac{t}{s}) |x|} .$$

If $E = S^{n-1}$, the same modulus is obtained with the family of paths joining the spheres $S^{n-1}(s)$ and $S^{n-1}(t)$ in the ring $\bar{B}^n(t) \setminus B^n(s)$. Here the ring can be replaced also by R^n.

A condenser in M is a pair $E = (A,C)$ where $A \subset M$ is open and $C \subset A$ compact. The conformal capacity or (n-)capacity of E is

$$\text{(5.3)} \qquad \operatorname{cap} E = \inf_u \int_A |\nabla u|^n dm$$

where the infimum is taken over all $u \in C_0^\infty(A)$ with $u|C = 1$ and where ∇u is the gradient of u. The value in (5.3) remains the same if u varies in $C(A) \cap W_{n,0}^1(A)$ where $W_{n,0}^1(A)$ is the closure with respect to $W_n^1(A)$ of the space $C_0^\infty(A)$. The connection to moduli of path families is expressed in the following

5.4 Proposition [Zi]. *If* $E = (A,C)$ *is a condenser with* $A \subset\subset M$, *then* $\operatorname{cap} E = M(\Gamma_E)$ *where* Γ_E *is the family of paths in* $\bar{A}$ *connecting* C *to* ∂A.

The inequality $M(\Gamma_E) \leq \operatorname{cap} E$ follows because if u is admissible for E, then $\rho = |\nabla u|$ is in $F(\Gamma_E)$. The other direction is more technical. The idea is that given ρ in $F(\Gamma_E)$ one defines $\rho_k = \min(\rho,k)$, $k = 1,2,\ldots,$ and $u_k : A \longrightarrow R^1$ by

$$u_k(x) = \inf_\alpha \int_\alpha \rho_k ds$$

where α runs over all rectifiable paths joining x to ∂A_k with $A_1 \subset A_2 \subset \ldots \subset A_k \subset\subset A$ and the sets A_k exhaust A. One can show that $|\nabla u_k| \leq \rho_k$ a.e. and that

$$\liminf_{k\to\infty} d_k \geq 1$$

where $d_k = \min\{u_k(x) : x \in C\}$. Then u_k/d_k is admissible for E, and as $k \to \infty$, we get $\operatorname{cap} E \leq M(\Gamma_E)$.

If $f: M \longrightarrow N$ is qc, then the modulus is a quasi-invariant, more precisely,

(5.5) $\frac{1}{K_0(f)} M(\Gamma) \leq M(f\Gamma) \leq K_I(f) M(\Gamma)$.

This is often used as a definition of qc maps as is the case in [V1]. In [V1] our definition for qc maps is obtained as a theorem.

For qr maps the right hand inequality is true. It is the Poleckii's theorem and it is presented in 5.9. On the other hand, the left hand inequality is not true in general for qr maps. A simple example to show this is the following.

Take $f: R^2 \longrightarrow R^2$ to be the exponential function $f(z) = e^z$ and let Γ be the family of line segments joining $(0, x_2)$ to $(1, x_2)$ and $0 \leq x_2 < k2\pi$. Then $M(\Gamma) = k2\pi$ and $M(f\Gamma) = 2\pi$. Hence $M(\Gamma)/M(f\Gamma) = k$ and this is exactly the covering number of f in $|\Gamma|$. In this sense the following result is sharp.

5.6. <u>Theorem</u> [MRV1,3.2] <u>Let</u> $f: M \longrightarrow N$ <u>be a nonconstant qr map</u>, <u>let</u> $A \subset M$ <u>be a Borel set with</u> $N(f,A) = \sup_y N(y,f,A) < \infty$ (<u>recall</u> $N(y,f,A) = \text{card } A \cap f^{-1}(y)$), <u>let</u> Γ <u>be a path family in</u> A. <u>Then</u>

$$M(\Gamma) \leq K_0(f) N(f,A) M(f\Gamma).$$

<u>Outline of proof.</u> Let $\rho' \in F(f\Gamma)$ and set

$$\rho(x) = \rho'(f(x)) L(x,f), \quad x \in A,$$

$$\rho(x) = 0, \quad x \in M \setminus A.$$

Here (d is the distance)

$$L(x,f) = \limsup_{d(x,y)\to 0} \frac{d(f(x),f(y))}{d(x,y)} .$$

If $\gamma \in \Gamma$ is locally rectifiable and f is locally absolutely continuous on γ (i.e. $f\circ\gamma^0$ is locally absolutely continuous where γ^0 is γ parametrized by arc length), then (see [V1,5.3])

$$\int_\gamma \rho ds \geq \int_{f\circ\gamma} \rho' ds \geq 1.$$

Hence if $\Gamma_0 = \{\gamma \in \Gamma : \gamma$ is locally rectifiable and f is locally absolutely continuous on $\gamma\}$, then $\rho \in F(\Gamma_0)$. The essential technical step is the following result.

5.7 Fuglede's theorem (see [V1,28.2]) $M(\Gamma) = M(\Gamma_0)$.

With 4.6(4) we get then finally

$$M(\Gamma) = M(\Gamma_0) \leq \int_M \rho^n dm = \int_A \rho'(f(x))^n L(x,f)^n dx$$

$$\leq K_0(f) \int_M (\rho'\circ f)^n J_f dm = K_0(f) \int_N \rho'(y)^n N(y,f,A)dy.$$

Remarks 1. The last inequality

$$(5.8) \qquad M(\Gamma) \leq K_0(f) \int_N \rho'(y)^n N(y,f,A)dy$$

is often more useful than 5.6. In fact, we have an occasion to use it in the proof of 7.3.

2. In the proof of 5.6 the function ρ is the density of a conformal metric which majorizes the metric obtained by pulling back ρ'.

Much more important and also more difficult to prove is the following Poleckii's inequality.

5.9 <u>Theorem</u> [Po] <u>If</u> $f:M \longrightarrow N$ <u>is a nonconstant</u> qr <u>map</u> <u>and</u> Γ <u>a path family in</u> M, <u>then</u>

$$M(f\Gamma) \leqq K_I(f)M(\Gamma).$$

<u>Outline of proof.</u> Let $\rho \in F(\Gamma)$ and set

$$\sigma(x) = \frac{\rho(x)}{\ell(f'(x))} \chi_E(x)$$

where $E = \{x \in M: f$ is differentiable at x and $J_f(x) > 0\}$ and

$$\ell(f'(x)) = \inf_{|h|=1} |f'(x)h| .$$

Then set

$$\rho'(y) = \sup\{\sigma(x): x \in f^{-1}(y)\}, \quad y \in fM,$$
$$\rho'(y) = 0, \qquad y \in N \setminus fM.$$

With this definition ρ' majorizes the various pushed down metrics.

Next we choose a sequence of Borel functions $\rho_j': N \longrightarrow [0,\infty[$ such that $\rho_j' \nearrow \rho'$ and $0 < \rho_j'(y) < \rho'(y)$ if

$\rho'(y) > 0$. Set $P_j = \{x \in M : \sigma(x) \geq \rho_j' \circ f(x)\}$. Then $N(y,f,P_j) \geq 1$ for $y \in fM$, and we get

$$\int_N \rho'^n dm \leftarrow \int_N \rho_j'^n dm \leq \int_N \rho_j'(y)^n N(y,f,P_j)\,dy = \int_{P_j} (\rho_j' \circ f)^n J_f dm$$

$$\leq \int_{P_j} \sigma^n J_f dm \leq K_I(f) \int_M \rho^n dm$$

where we have used 4.6.

It remains to show that ρ' is admissible for a family $f\Gamma_0$ such that $M(f\Gamma) = M(f\Gamma_0)$. This is the technically hard part and we shall only give the result without proof known as Poleckii's Lemma.

Let $\gamma : [a,b] \longrightarrow M$ be a path such that $f \circ \gamma$ is rectifiable. Let $(f\circ\gamma)^0$ be the latter parametrized by arc length and $s : [a,b] \longrightarrow [0, \ell(f\circ\gamma)]$ the length function, i.e. $f\circ\gamma = (f\circ\gamma)^0 \circ s$. Because f is discrete, there is a path $\gamma^* : [0,\ell(f\circ\gamma)] \longrightarrow M$ such that $\gamma = \gamma^* \circ s$. The path γ^* is called the f-representation of γ. We say that f is <u>absolutely precontinuous</u> on γ is γ^* is absolutely continuous. Note that if f is injective, then f is absolutely precontinuous on γ if and only if f^{-1} is absolutely continuous on $f\circ\gamma$.

5.10. <u>Poleckii's Lemma.</u> [Po, Lemma 6]. <u>If</u> $\Gamma_1 \subset \Gamma$ <u>is the subfamily of paths for which</u> $f\circ\gamma$ <u>is not locally rectifiable or</u> f <u>is not locally absolutely precontinuous on</u> γ, <u>then</u>

$M(f\Gamma_1) = 0$.

To continue the proof of 5.9 set $\Gamma_0 = \Gamma \setminus \Gamma_1$. Then (see [V1,5.3])

$$1 \leq \int_\gamma \rho ds \leq \int_{f\circ\gamma} \rho' ds \quad \text{if} \quad \gamma \in \Gamma_0 ,$$

hence $\rho' \in F(\Gamma_0)$. Poleckii's lemma gives $M(f\Gamma_0) = M(f\Gamma)$ and the proof is finished.

Remark. The estimate $N(y,f,P_j) \geq 1$ apearing in the proof is often crude. A sharper inequality which takes this into account is the following, known as Väisälä's inequality.

5.11. Theorem [V2,3.1] *Let* $f:M \longrightarrow N$ *be a nonconstant qr map,* Γ *a path family in* N, Γ^* *a path family in* M, *and* m *a positive integer. Suppose that for each* $\beta \in \Gamma$ *there are paths* $\alpha_1,\ldots,\alpha_m$ *in* Γ^* *such that each* α_i *is a (partial) lift of* β *and*

$$\text{card } \{j : \alpha_j(t) = x\} \leq i(x,f) \quad \textit{for all} \quad x,t .$$

Then

$$M(\Gamma) \leq \frac{K_I(f)}{m} M(\Gamma^*).$$

The proof is similar to the one of 5.9. In particular Poleckii's Lemma is used in the same way. The essential change is in the definition of the function ρ' which this time is defined by

$$\rho'(y) = \frac{1}{m} \sup_B \sum_{x \in B} \sigma(x)$$

where B runs over all subsets of $f^{-1}(y)$ with $\operatorname{card} B \leq m$.

Remark. The proofs of 5.9 and 5.10 in [Po] and the proof of 5.11 in [V2] are given in the Euclidean case. Only minor changes are required for the manifold case; for this, see [MR]. The complete proofs will appear also in the forthcoming book [Ri8].

We finish this section by some simple applications. Väisälä's inequality will be applied in Section 7.

Let M be a Riemannian n-manifold and $F \subset M$ compact, $F \neq M$. We say that F has capacity zero, denoted by $\operatorname{cap} F = 0$, if $M(\Gamma) = 0$ where Γ is the family of paths $\gamma: [a,b[\longrightarrow M \setminus F$ with $\gamma(t) \longrightarrow F$ as $t \longrightarrow b$. A set $E \subset M$ has capacity zero if $\operatorname{cap} F = 0$ for all compact $F \subset E$. If E does not have capacity zero, we write $\operatorname{cap} E > 0$. Note that $\operatorname{cap} E$ has no meaning as a number in this case. Zero capacity sets are metrically thin the following sense.

5.12 Proposition [Re1]. *If $E \subset M$ is a Borel set and $\operatorname{cap} E = 0$, then the Hausdorff dimension $\dim_H E$ is zero.*

At an early stage the following result on the set of omitted values was proved both in [MRV2,4.4] and [Re4].

5.13 Theorem. *If $f: R^n \longrightarrow R^n$ is a nonconstant qr map, then $\operatorname{cap}(R^n \setminus fR^n) = 0$.*

<u>Proof.</u> Let $E = R^n \setminus fR^n$ and suppose $\operatorname{cap} F > 0$ for some compact $F \subset E$. Let Γ be the family of paths $\gamma:[a,b] \longrightarrow R^n$ such that $\gamma(a) \in f\overline{B}^n$ and $\gamma(b) \in F$. Since $f\overline{B}^n$ is a continuum, it follows from properties of the moduli and $\operatorname{cap} F > 0$ that $M(\Gamma) > 0$. Since f is discrete and open, every path $\gamma \in \Gamma$ has a maximal lift α starting in $\overline{B}^n$ (see [Ri1]) and such a lift α must tend to ∞ because $\gamma(b) \in F$. Let Γ^* be the family of such lifts. Then $M(\Gamma^*) = 0$ which can be proved for example by application of Example 5.2(b). Since every path $\gamma \in \Gamma$ contains a subpath in $f\Gamma^*$, we have $M(f\Gamma^*) \geqq M(\Gamma)$. By Poleckii's inequality we then are led to the contradiction

$$0 = K_I(f)M(\Gamma^*) \geqq M(f\Gamma^*) \geqq M(\Gamma).$$

<u>Remark.</u> Exactly the same proof leads to the following more general result. Let $f:M \longrightarrow N$ be a nonconstant qr map, let Γ (resp. Γ') be the family of paths in N (resp. M) connecting a continuum to ∞. We assume that M and N are noncompact. Suppose $M(\Gamma') = 0$. Then also $M(\Gamma) = 0$.

From the inequalities for moduli of path families one easily obtains inequalities for capacities of condensers. Historically the latter where proved first [MRV1]. Let $f:M \longrightarrow N$ be a nonconstant qr map and (A,C) a condenser in N. From Poleckii's inequality we get the inequality

$$(5.14) \qquad \operatorname{cap}(fA,fC) \leqq K_I(f)\operatorname{cap}(A,C).$$

To prove (5.14) we may assume $A \subset\subset M$. Let Γ be the family of paths in $\overline{fA}$ connecting fC to ∂fA, and let Γ^* be the family of maximal lifts in $\overline{A}$ starting in C. Then all paths in Γ^* connect C and ∂A and by 5.4 and 5.9,

$$\operatorname{cap}(fA,fC) = M(\Gamma) \leq M(f\Gamma^*) \leq K_I(f)M(\Gamma^*) \leq K_I(f)\operatorname{cap}(A,C).$$

An inequality corresponding to 5.6 is true for special, so called normal condensers, see [MRV1].

To state a substitute for Schwarz's Lemma we need some facts about the Grötzsch condenser $(B^n,[0,re_1])$. Its capacity is denoted by $\nu_n(r)$. For $K \geq 1$ we set $\varphi(n,K,r) = \nu_n^{-1}(K\nu_n(r))$. Then $\varphi(n,K,r)$ is continuous and strictly increasing in r with boundary values $\varphi(n,K,0) = 0$, $\varphi(n,K,1) = 1$. Moreover, it satisfies [Ge1,p.518]

$$\varphi(n,K,r) \leq \lambda_n r^{K^{1/(n-1)}}$$

where the constant $\lambda_n \geq 1$ depends only on n.

5.15. <u>Theorem</u> [MRV2,3.1]. <u>Suppose</u> $f:B^n \longrightarrow B^n$ <u>is</u> qr <u>and</u> $f(0) = 0$. <u>Then</u>

$$|f(x)| \leq \varphi(n,K_I(f),|x|).$$

<u>Proof.</u> By (5.14), $\operatorname{cap}(fB^n,f[0,x]) \leq K_I(f)\nu_n(|x|)$. By the symmetrization result for condensers $\nu_n(|f(x)|) = \operatorname{cap}(B^n,[0,f(x)]) \leq \operatorname{cap}(fB^n,f[0,x])$, and the result follows.

As in the qc case the moduli inequalities are convenient for characterization of qr maps too. An example of such a result is the following.

5.16. Theorem. *Let* M *and* N *be Riemannian n-manifolds and* $f: M \longrightarrow N$ *a sensepreserving, discrete and open map. Then the following two conditions are equivalent:*

(1) f *is* qr *and* $K_I(f) \leq K$.

(2) $M(f\Gamma) \leq KM(\Gamma)$ *for all path families* Γ *in* M.

The direction (1) ⇒ (2) is Poleckii's inequality. For the proof of (2) ⇒ (1) we refer to [MRV1,7.1].

5.17. Corollary. *If* f *and* g *are* qr *and* $g \circ f$ *is defined, then* $g \circ f$ *is* qr. *Moreover,*

$$K_I(g \circ f) \leq K_I(g) K_I(f).$$

Proof. We may assume that $g \circ f$ is nonconstant. Then f, g, and $g \circ f$ are discrete and open. Theorem 5.16 gives first

$$M((g \circ f)\Gamma) \leq K_I(g) M(f\Gamma) \leq K_I(g) K_I(f) M(\Gamma)$$

for all path families Γ in the domain of f. Then we get that $g \circ f$ is qr and $K_I(g \circ f) \leq K_I(g) K_I(f)$.

As a final application we shall prove for a nonconstant qr map $f: M \longrightarrow N$ that $J_f(x) > 0$ a.e. and $m(B_f) = 0$ (Theorem 4.6 (5),(6)) by using Poleckii's lemma and 5.6. This proof arrangement is due to Pesonen [Pe2].

<u>Proof of</u> 4.6 (5),(6). Suppose $J_f(x) = 0$ in a set E of positive measure. We may assume that E is a Borel set contained in a closed cube Q and that for all $x \in E$ f is differentiable at x and $f'(x) = 0$. Let Γ be the family of paths γ with $|\gamma| = L \cap Q$, where L is a line parallel to e_1, and

$$\int_\gamma \chi_E ds > 0.$$

Fubini's theorem with 5.2(a) implies $M(\Gamma) > 0$. The inequality 5.6 gives

$$0 < \frac{M(\Gamma)}{K_0(f)N(f,Q)} \leq M(f\Gamma).$$

Now Poleckii's Lemma says that there exists $\gamma \in \Gamma$ such that the f-representation γ^* is absolutely continuous, and therefore

$$0 < \int_\gamma \chi_E ds = \int_0^{\ell(f\circ\gamma)} (\chi_E \circ \gamma^*)|\gamma^{*\prime}| dm_1 = \int_{\gamma^{*-1}E} |\gamma^{*\prime}| dm_1 .$$

Hence $m_1(\gamma^{*-1}E) > 0$. On the other hand, for a.e. $t \in \gamma^{*-1}E$ we have

$$1 = |(f\circ\gamma)^{0\prime}(t)| = |f'(\gamma(t))\gamma^{*\prime}(t)| = 0$$

which is a contradiction. We have showed that $J_f(x) > 0$ a.e.

Since f is differentiable a.e., it follows that $J_f(x) = 0$ a.e. in B_f. But $J_f(x) > 0$ a.e., hence $m(B_f) = 0$.

6. Locally homeomorphic quasiregular maps

In this section we shall prove the main result for locally homeomorphic qr maps in dimensions $n \geqq 3$, namely Zorič's theorem. In fact, we shall prove the following more general result formulated for maps of a ball.

6.1. Theorem [MRV3,2.3] *Suppose* $f: B^n \longrightarrow R^n$ *is* K-qr, *locally homeomorphic, and* $n \geqq 3$. *Then there exists* $\psi = \psi(n,K) > 0$ *such that* $f|B^n(\psi)$ *is injective.*

Zorič's theorem follows as a corollary:

6.2. Corollary [Zo] *Suppose* $f: R^n \longrightarrow R^n$ *is* qr, *locally homeomorphic, and* $n \geqq 3$. *Then* f *is* qc.

For the proof of 6.1 we need some topological lemmas which we state without proofs.

6.3. Lemma [MRV3,2.2]. *Let* G *be a domain in* R^n, $f: G \longrightarrow S^n$ *locally homeomorphic*, $Q \subset S^n$ *simply connected and locally pathwise connected*, *and* P *a component of* $f^{-1}Q$ *such that* $\overline{P}$ *is a compact subset of* G. *Then* f *maps* P *homeomorphically onto* Q. *If moreover*, Q *is relatively locally connected, i.e. every point in* $\overline{Q}$ *has arbitrarily small neighborhoods* U *such that* $U \cap Q$ *is connected, then* f *maps* $\overline{P}$ *homeomorphically onto* $\overline{Q}$.

6.4. Lemma [Zo,p.422]. *Let* $f: G \longrightarrow S^n$ *be as in* 6.3, *let* $A, B \subset G$ *be sets such that* $f|A$ *and* $f|B$ *are injective,*

$A \cap B \neq \phi$, <u>and</u> $fA \cap fB$ <u>is connected</u>. <u>Then</u> f <u>is injective in</u> $A \cup B$.

<u>Proof of 6.1.</u> We may assume $f(0) = 0$. Let $U(r)$ be the 0-component of $f^{-1}B^n(r)$ and set $r_0 = \sup\{r:\overline{U}(r) \subset B^n\}$. Fix $0 < r < r_0$ and write $U = U(r)$. Lemma 6.3 implies that $f|\overline{U}:\overline{U} \longrightarrow \overline{B}^n(r)$ is a homeomorphism. Set $\ell^* = \min\{|z|:z \in \partial U\}$, $L^* = \max\{|z|:z \in \partial U\}$, $\ell = \min\{|y|:y \in fS^{n-1}(\ell^*)\}$. It suffices to find a lower bound of ℓ^*.

In what follows we may assume $\ell < r$. Then $A = U \setminus \overline{U}(\ell)$ is a ring and both components of ∂A meet $S^{n-1}(\ell^*)$. It follows that the capacity satisfies $\text{cap}\,(U,\overline{U}(\ell)) \geqq a_n > 0$ [V1,11.7]. Since f is qc in U, we also have

$$a_n \leqq \text{cap}\,(U,\overline{U}(\ell)) \leqq K\,\text{cap}\,(fU,f\overline{U}(\ell)) = \frac{K\,\omega_{n-1}}{(\ell n\,\frac{r}{\ell})^{n-1}}$$

where ω_{n-1} is the $(n-1)$-measure of the unit sphere S^{n-1}. This gives an inequality $r/\ell \leqq \alpha(n,K)$ which holds also in the case $\ell = r$.

Next fix a point x_0 in ∂U such that $|x_0| = L^*$ and set $y_0 = f(x_0)$. For $r < t < r + \ell$ we consider caps $C(t,\varphi) = \{y \in R^n:|y - y_0| = t,\ (y_0 - y)\cdot y_0 > rt\cos\varphi\}$ of the sphere $S^{n-1}(y_0,t)$ with axis the line L through 0 and y_0 and with angle φ, $0 < \varphi \leqq \pi$. Let z_t be the point $(r - t)r^{-1}y_0$ in the line segment $J = \{sy_0:-\ell/r < s < 0\} \subset L$, let z_t^* be the unique point in $U \cap f^{-1}(z_t)$, and let $C^*(t,\varphi)$ be the z_t^*-component of

$f^{-1}C(t,\varphi)$ (see Figure).

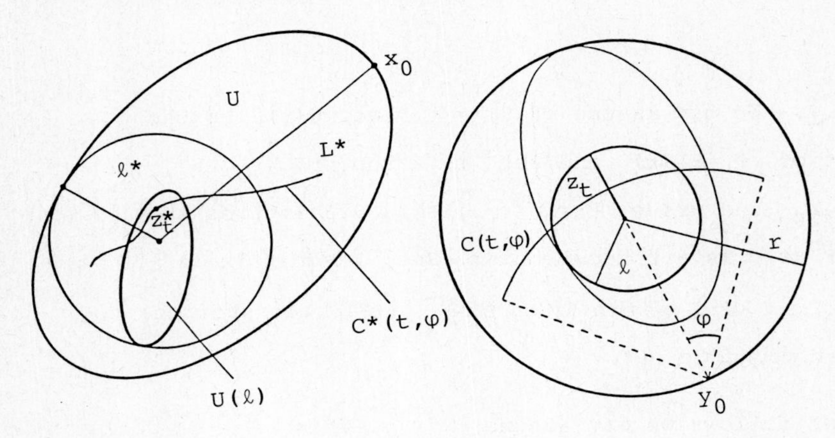

Now let φ_t be the supremum of all $\varphi \in]0,\pi]$ for which f maps $C^*(t,\varphi)$ homeomorphically onto $C(t,\varphi)$. We claim that $C^*(t,\varphi_t) \cap S^{n-1}(L^*) \neq \phi$ for $r < t < r + \ell$. Suppose this is not true. Then for some $t \in]r,r+\ell[$ $C^*(t,\varphi_t) \subset B^n(L^*)$. Lemma 6.3 implies that f maps $\overline{C}^*(t,\varphi_t)$ homeomorphically onto $\overline{C}(t,\varphi_t)$. Note that for $n = 2$ the proof breaks down here because $C(t,\pi)$ is not relatively locally connected. Since f is locally homeomorphic, f is injective in a neighborhood of $\overline{C}^*(t,\varphi_t)$. It follows that the only possibility is $\varphi_t = \pi$, hence $\overline{C}^*(t,\varphi_t) = \overline{C}^*(t,\pi)$ is a topological $(n-1)$-sphere. Let $D \subset B^n(L^*)$ be the bounded component of $R^n \setminus \overline{C}^*(t,\varphi_t)$. Then $\partial fD \subset S^{n-1}(y_0,t)$, hence $fD = B^n(y_0,t)$. Again by Lemma 6.3 f map $\overline{D}$ homeomorphically onto $\overline{B}^n(y_0,t)$. The point z_t^* is in $\overline{D} \cap \overline{U}$, so $\overline{D} \cap \overline{U} \neq \phi$. Also $\overline{B}^n(y_0,t) \cap \overline{B}^n(r) = f\overline{D} \cap f\overline{U}$ is connected. Lemma 6.4 implies then that f in injective in $\overline{D} \cup \overline{U}$. But this gives $x_0 \in D$ because $y_0 \in fD$ which is a contradiction. We have proved that $C^*(t,\varphi_t) \cap S^{n-1}(L^*) \neq \phi$ for $r < t < r + \ell$.

For $r < t < r + \ell$ choose $x_t^* \in C^*(t,\varphi_t) \cap S^{n-1}(L^*)$ and let Γ_t be the family of paths in $C^*(t,\varphi_t)$ joining z_t^* and x_t^*. Let Γ be the union of the families Γ_t, $r < t < r + \ell$. Since $|z_t^*| \leq \ell^*$ and $|x_t^*| = L^*$, we have

$$M(\Gamma) \leq \frac{\omega_{n-1}}{(\ln \frac{L^*}{\ell^*})^{n-1}} .$$

On the other hand, $M(f\Gamma) \geq b_n \ln(1 + \ell/r)$ [V1,10.12]. Combined with $M(f\Gamma) \leq KM(\Gamma)$ these equalities give a bound $\ell^* \geq L^*\psi(n,K)$. The number L^* tends to 1 as $r \longrightarrow r_0$. The theorem is proved.

The results 6.1 and 6.2 are not true for $n = 2$ which can be seen by means of the exponential function $z \longmapsto e^z$.

Zorič's theorem implies some other interesting results. For example one can use it to show that a holomorphic map $f:\mathbb{C}^k \longrightarrow \mathbb{C}^k, k > 1$ which is qr as a map $R^{2k} \longrightarrow R^{2k}$, is necessarily affine [MaR]. This result shows in a sense that the theory of holomorphic maps in complex dimension > 1 and the qr theory are generalizations of the one complex variable theory in completely different directions.

7. The analogue of Picard's theorem

In 1880 Picard proved the well-known result that a complex analytic function $f:R^2 \longrightarrow R^2 \setminus \{a_1,a_2\}$ must be constant. The same result is a also true for planar qr maps which can immediately be seen from the decomposition $f = g \circ f$ of a qr map f where h is qc and g analytic.

For the smooth case this was proved by Grötsch in 1928.

In 1967 Zorič [Zo] posed the question whether a Picard type theorem is true also for $n \geq 3$. For a long time it was conjectured that a Picard's theorem is true in the same form as in the plane. In 1980 a Picard type theorem appeared in the following form.

7.1. Theorem [Ri3]. *There exists an integer* $q = q(n,K)$ *such that every* K-qr *map* $f: R^n \longrightarrow R^n \setminus \{a_1, \ldots, a_q\}$ *is constant.*

It is now known for $n = 3$ that the conjecture mentioned above is false and, in fact, Theorem 7.1 is qualitatively best possible. The complementing result is as follows.

7.2. Theorem [Ri7]. *For every positive integer* p *there exists a nonconstant* qr *map* $f: R^3 \longrightarrow R^3$ *omitting* p *points.*

Apart from [Ri3] there exist two other proofs of 7.1 in the articles [Ri5] and [Ri6]. The proof in [Ri5] is a simplification of the proof of a more general result, namely a defect relation in [Ri4]. We shall here give a fairly detailed presentation of that proof. On the other hand, the proof of 7.2 is very technical and we shall here only make some remarks about it later.

The essential tool in our proof for 7.1 is a certain comparison lemma for coverings over spheres. In the proof of the lemma we need Väisälä's modulus inequality (Theorem 5.11) together with a path lifting result. For the formulation we need some notation.

Let $f:R^n \longrightarrow R^n$ be a nonconstant qr map and write $K_I = K_I(f)$. For $E \subset R^n$ such that $\overline{E} \subset R^n$ is compact and for $y \in R^n$ set

$$n(E,y) = \sum_{x \in f^{-1}(y) \cap E} i(x,f)$$

where we recall the notation $i(x,f)$ for the local index from 3.1. If E is a ball $\overline{B}^n(r)$, we write $n(r,y) = n(\overline{B}^n(r),y)$ and this is often called the <u>counting function.</u> If $Y \subset R^n$ is an $(n-1)$-sphere, we let $\nu(E,Y)$ be the average of $n(E,y)$ over Y with respect to the $(n-1)$-measure. If $E = \overline{B}^n(r)$ and $Y = S^{n-1}(t)$, we write $\nu(r,t) = \nu(E,Y)$. In the following we also write S for the unit sphere S^{n-1}. Hence

$$\nu(r,t) = \frac{1}{\omega_{n-1}} \int_S n(r,ty)\,dy .$$

7.3. <u>Comparison lemma.</u> <u>If</u> $\theta > 1$ <u>and</u> $r,s,t > 0$, <u>then</u>

$$\nu(\theta r,t) \geqq \nu(r,s) - \frac{K_I |\ln(t/s)|^{n-1}}{(\ln\theta)^{n-1}} .$$

<u>Proof.</u> We may assume $s < t$. For $m = 1,2,\ldots$ set

$$E_m = \{y \in S : n(\theta r,ty) = n(r,sy) - m\} ,$$

$$E = \bigcup_m E_m .$$

Then

$$(1) \qquad \int_S n(\theta r,ty)\,dy = \int_{S \setminus E} n(\theta r,ty)\,dy + \int_E n(\theta r,ty)\,dy$$

$$\geq \int_{S\setminus E} n(r,sy)\,dy + \sum_m \int_{E_m} (n(r,sy) - m)\,dy$$

$$= \int_S n(r,sy)\,dy - \sum_m m\, m_{n-1}(E_m).$$

For $y \in S$ let $\beta_y : [s,t] \longrightarrow R^n$ be the path defined by $\beta_y(u) = uy$ and set $\Gamma_m = \{\beta_y : y \in E_m\}$. Let $y \in E_m$. From [Ri1] we obtain the following path lifting result: There exists a sequence $\alpha_1,\dots,\alpha_k$, $k = n(r,sy)$, of maximal $f|B^n(\theta r)$-lifts of β_y starting at points in $f^{-1}(y) \cap \overline{B}^n(r)$ such that

$$\text{card}\,\{j : \alpha_j(t) = x\} \leq i(x,f)$$

for all x and t (which make sense). Since $y \in E_m$, at least m of $\alpha_1,\dots,\alpha_k$ end in $\partial B^n(\theta r)$. Call the family of those lifts when y runs over E_m by Γ_m^*. Väisälä's inequality 5.11 gives now

$$(2) \qquad M(\Gamma_m) \leq \frac{K_I}{m} M(\Gamma_m^*) \ .$$

Since the path families Γ_m are separate, so are the families Γ_m^*. By properties of the modulus

$$(3) \qquad \sum_m M(\Gamma_m^*) = M(\bigcup_m \Gamma_m^*) \leq \frac{\omega_{n-1}}{(\ln\theta)^{n-1}} \ .$$

On the other hand,

$$(4) \qquad M(\Gamma_m) = \frac{m_{n-1}(E_m)}{(\ln\frac{t}{s})^{n-1}} \ .$$

From (2)-(4) we get

$$\sum_m m\, m_{n-1}(E_m) \leq \frac{K_I \omega_{n-1} (\ln(t/s))^{n-1}}{(\ln\theta)^{n-1}} .$$

With (1) this gives the claim in the lemma.

Remark. Lemma 7.3 is in a slightly weaker form in [Ri2,4.1]. The present form was proved essentially with the same proof as in [Ri2] by Pesonen [Pe1] and Hinkkanen (independently).

7.4. Lemma. *If* ∞ *is an essential singularity of* f, *i.e.* f *has no limit in* $\overline{R}^n$ *at* ∞ *then*

$$\lim_{r\to\infty} \nu(r,s) = \infty \quad \textit{for all} \quad s > 0.$$

Proof. By 7.3 we may assume $s = 1$. By slight modification of the proof of 5.13 we can show that $\operatorname{cap}(R^n \setminus f(R^n \setminus B^n(r))) = 0$ for all $r > 0$. This is turn implies the existence of a Borel set $E \subset R^n$ of zero capacity such that for all $r > 0$,

$$N(y,f,R^n \setminus B^n(r)) = \infty, \quad y \in R^n \setminus E.$$

Set $F_k(r) = \{y \in S : n(r,y) \geq k\}$, $k = 1,2,\ldots$ Then

$$\lim_{r\to\infty} \nu(r,1) \geq \lim_{j\to\infty} \frac{1}{\omega_{n-1}} \int_{F_k(j)} k\,dy = \frac{k}{\omega_{n-1}} \lim_{j\to\infty} m_{n-1}(F_k(j))$$

$$\geq \frac{k}{\omega_{n-1}} m_{n-1}(S \setminus H)$$

where H is a Borel set of zero capacity. But $m_{n-1}(H) = 0$ by 5.12, and the Lemma follows.

Remark. In the proof we used only the properties that ν is an increasing positive function and that $\alpha > 0$.

Proof of 7.1. We assume that $f: R^n \longrightarrow R^n \setminus \{a_1,\ldots,a_q\}$ is a nonconstant K-qr map and $a_1,\ldots,a_q$ are distinct. We may assume that the omitted points a_j lie all in $B^n(1/2)$. Set

$$\sigma_0 = \frac{1}{4} \min_{j \neq k} |a_j - a_k| .$$

Fix $s \in [\kappa, \infty[\setminus E$ where $\kappa \geq r_0''$ is a number chosen sufficiently large later. With the notation of Lemma 7.5 set $d_0 = s' - s$. We decompose $\overline{B}^n(s)$ into disjoint Borel sets $U_i, i \in I = \{1,\ldots,p\}$, such that for some K_0 - qc map φ_i of R^n onto itself, $\varphi_i^{-1} B^n(1/2) \subset U_i \subset \varphi_i^{-1} \overline{B}^n(1/2)$, and $0 < \alpha_n \leq \rho(U_i) \leq \beta_n < \infty$, where K_0, α_n, β_n depend only on n and $\rho(U_i)$ is the diameter of U_i in the hyperbolic metric of the ball $B^n(s')$ with density $1/(s'^2 - |x|^2)$. An elementary calculation gives an estimate

$$p \leq b_1 \left(\frac{s}{d_0}\right)^{n-1} = b_1 \nu(s)^{1/2} . \tag{1}$$

Here we shall denote by b_k positive constants that depend only on n and by c_k positive constants that depend only on n and K, $k = 0,1,\ldots$ Write

$$Y_i = \varphi_i^{-1} B^n, \quad Z_i = \varphi_i^{-1} \overline{B}^n(3/2), \quad W_i = \varphi_i^{-1} \overline{B}^n(2).$$

We may assume that $W_i \subset \overline{B}^n(s')$ for all i and that the sets W_i do not overlap more than b_2 times.

Note that if ∞ is not an essential singularity of f, then f can be extended to a qr map $f:\overline{R}^n \longrightarrow \overline{R}^n$ and $\nu(r,s) \longrightarrow k$ = degree of f as $r \longrightarrow \infty$.

7.5. Lemma. Set $\nu(r) = \nu(r,1)$. There exists a set $E \subset [1,\infty[$ of finite logarithmic measure, i.e.

$$\int_E \frac{dr}{r} < \infty,$$

such that

$$\nu(r') \leqq 2\nu(r) \quad \text{for} \quad r \in [1,\infty[\setminus E,$$

where $r' = r + r/\nu(r)^{\alpha}$, $\alpha = 2^{-1}(n-1)^{-1}$.

Proof. The idea of the argument goes back to E. Borel. Let $r_0'' \geqq 1$ be such that $\nu(r_0'') \geqq 1$ and set $F = \{r \in\ ]r_0'',\infty[\ : \nu(r') > 2\nu(r)\}$. We define inductively $r_0'' \leqq r_1 < r_1'' \leqq r_2 < r_2'' \leqq \ldots$ (possible a finite sequence) by

$$r_k = \inf\{r \in F : r > r_{k-1}''\},$$

$$r_k'' = r_k + \frac{2r_k}{\nu(r_k)^{\alpha}} .$$

Let E_1 be the union of the intervals $[r_k, r_k'']$, $k \geqq 1$. We claim that $E = [1,r_0''] \cup E_1$ satisfies our requirements. We observe that $F \subset E_1$ (If $F = \phi$, we define $E_1 = \phi$). We have $\nu(r_{k+1}) \geqq \nu(r_k'') \geqq 2\nu(r_k)$. The Lemma follows then from

$$\int_{E_1} \frac{dr}{r} \leqq \sum_{k\geqq 1} \frac{r_k''-r_k}{r_k} = \sum_{k\geqq 1} \frac{2}{\nu(r_k)^{\alpha}} < \infty .$$

Recall the notation $\nu(E,Y)$ before 7.3 for the average cover from E over an $(n-1)$-sphere Y. Set $\nu(E) = \nu(E,S^{n-1})$. We shall next find $i \in I$ such that $\nu(W_i)/\nu(U_i)$ has a bound depending only on n and such that $\nu(U_i)$ is sufficiently large. This will be the essential point where we use the fact that f is a map of R^n. For this set

$$I_0 = \{i \in I : \nu(W_i) \geqq 3b_2\, \nu(U_i),$$

$$I_1 = \{i \in I : \nu(U_i) \leqq \nu(s)^{1/4}\}.$$

We claim that $I \setminus (I_0 \cup I_1) \neq \phi$ if κ is sufficently large.

We first observe that ∞ is an essential singularity of f because otherwise f would extend to a map of $\overline{R}^n$ onto a proper compact subset of $\overline{R}^n$ which would contradict 5.13. Hence by 7.4, $\nu(r) \longrightarrow \infty$ as $r \longrightarrow \infty$. By the choice of s and by (1), we obtain for sufficently large κ the estimate

$$\sum_{i \in I_0 \cup I_1} \nu(U_i) \leqq \sum_{i \in I_0} \nu(U_i) + \sum_{i \in I_1} \nu(U_i) \leqq \frac{1}{3b_2} \sum_{i \in I_0} \nu(W_i) + b_1 \nu(s)^{3/4}$$

$$\leqq \frac{1}{3}\nu(s') + b_1\nu(s)^{3/4} \leqq \frac{2}{3}\nu(s) + b_1\nu(s)^{3/4} \leqq \frac{3}{4}\nu(s) \qquad .$$

But

$$\sum_{i \in I} \nu(U_i) = \nu(s),$$

hence $I \setminus (I_0 \cup I_1) \neq \phi$.

Fix $i \in I \setminus (I_0 \cup I_1)$. Suppose that $a_j = 0$. We apply 7.3 to the map $g = f \circ \varphi_i^{-1}$ and get

$$\nu_g(1,\sigma) \geq \nu_g(1/2) - \frac{KK_0 (\ln \frac{1}{\sigma})^{n-1}}{(\ln 2)^{n-1}} \, , \quad 0 < \sigma < \sigma_0 \, ,$$

where we have indicated by subscript that the counting functions are with respect to the map g. Since $i \notin I_1$, we can for sufficiently large κ choose σ so that

$$(2) \qquad \nu(U_i) \leq \nu_g(1/2) = c_0 (\ln \frac{1}{\sigma})^{n-1} \lneqq 2c_0 (\ln \frac{\sigma_0}{\sigma})^{n-1}$$

where $c_0 = 2KK_0(\ln 2)^{1-n}$. Then $\nu_g(1,\sigma) > 0$ and hence $g^{-1}\overline{B}^n(a_j,\sigma) \cap \overline{B}^n \neq \phi$. By using an auxiliary qc map we get the same statement also if $a_j \neq 0$ with a larger c_0.

Let Γ_j be the family of paths in $B^n(3/2) \setminus \overline{B}^n$ joining $g^{-1}\overline{B}^n(a_j,\sigma)$ and some $g^{-1}\overline{B}^n(a_k,\sigma)$, $k \neq j$. It follows from the modulus estimate [V1,10.12] that $\max_j M(\Gamma_j) \geq a(n,q)$ where $a(n,q) \longrightarrow \infty$ as $q \longrightarrow \infty$. With some extra effort (for details, see [Ri4, p.188]) one can take $a(n,q) = b_3 q^{1/(n-1)}$. Note that since each a_k is omitted, each component of $g^{-1}\overline{B}^n(a_k,\sigma)$ tends to ∞. Fix now j such that

$$(3) \qquad M(\Gamma_j) \geq b_3 q^{1/(n-1)} \, .$$

Each path $g \circ \gamma$, $\gamma \in \Gamma_j$, connects $\overline{B}^n(a_j,\sigma)$ and some $\overline{B}^n(a_k,\sigma)$, $k \neq j$. Write $K_1 = KK_0$ and $M = \sigma_0/\sigma$. We may again assume $a_j = 0$. We define $\rho \in F(g\Gamma_j)$ by

$$\rho(y) = \frac{1}{(\ln M)|y|} \quad \text{if} \quad y \in B^n(\sigma_0) \setminus \overline{B}^n(\sigma),$$

$$\rho(y) = 0 \quad \text{elsewhere.}$$

The inequality (5.8) gives

$$(4) \qquad M(\Gamma_j) \leq K_1 \int_{R^n} \rho(y)^n n_g(3/2,y)\,dy = K_1 \int_{\sigma}^{\sigma_0}\left(\int_S \frac{n_g(3/2,tz)}{(\ln M)^n t}dz\right)dt$$

$$= \frac{K_1 \omega_{n-1}}{(\ln M)^n} \int_{\sigma}^{\sigma_0} \frac{\nu_g(3/2,t)}{t}\,dt.$$

Using the Comparison Lemma 7.3 again together with $i \notin I_0$ and (2) we get

$$\nu_g(3/2,t) \leq \nu_g(2) + \frac{K_1(\ln \frac{1}{t})^{n-1}}{(\ln \frac{4}{3})^{n-1}} \leq \nu(W_i) + c_1(\ln M)^{n-1}$$

$$\leq c_2(\ln M)^{n-1}.$$

With (4) this gives $M(\Gamma_j) \leq c_3$ and the theorem follows then from (3).

7.6. Remarks on the proof of 7.2. In Example 3.2 we saw how to construct a nontrivial qr map of R^3 omitting one point in R^3. The construction generalizes to R^n in a straightforward manner. On the other hand, already for $p = 2$ one can show that the map in Theorem 7.2 must satisfy hard requirements and this necessarily leads to a complicated construction. To describe this a little let us look at a given nonconstant qr map $f: R^3 \longrightarrow R^3 \setminus \{u_2, u_3\}$, $u_2 = -e_3/2$, $u_3 = e_3/2$. Write $u_1 = \infty$

and let U_1, U_2, U_3 be the components of $R^3 \setminus (S^2 \cup B^2 \cup \{u_2, u_3\})$ such that $u_j \in \overline{U}_j$, $j = 2, 3$. Set $W_j = f^{-1}U_j$. By arguments in value distribution, where the comparison Lemma 7.3 plays again an essential role, one can show that for each common boundary point x of W_j and W_k the third preimage W_ℓ must in a sense be near x. This condition alone makes the sets W_j very complicated when we approach ∞.

In the construction in [Ri7] also a fairly general 2-dimensional deformation theory for discrete open maps is needed. This is one of the essential features in the proof to keep the dilatation bounded.

In principle it seems to be possible to prove 7.2 also for $n \geqq 4$. However, the deformation theory and also some other parts should be modified considerably because in [Ri7] some properties of the plane are used essentially.

8. The general mapping problem

Our mapping problem is the following. Given two Riemannian n-manifolds M and N (recall that all manifolds are here assumed to be connected and oriented) does there exist a nonconstant qr map $f: M \longrightarrow N$? One of the most interesting problems of this type has been the problem of a Picard type theorem, namely the case $M = R^n$, $N = S^n \setminus \{a_1, \ldots, a_{q+1}\}$, considered in Section 7.

In general, not very much is known about the mapping problem. Let us first look at the case $N = S^n$. If M is given a triangulation such that for some $L \geq 1$ each n-simplex can be mapped by an L-bilipschitz map followed by a homothety onto a standard simplex, then there exists a nonconstant qr mapping $f: M \longrightarrow S^n$. We can construct f for example by performing a barycentric subdivision once and then using the Alexander's construction where every other n-simplex is mapped onto the upper hemisphere of S^n and every other onto the lower. Apart from this remark all results in this section concern the case $M = R^n$.

8.1 Isoperimetric inequalities. The connection of isoperimetric inequalities and capacity has been used by Gromov and Pansu to obtain results for the mapping problem, see [Gr2, Chapter VI], [Pa]. The idea goes back to Ahlfors' famous paper [A,p.188]. The proof of the following result can be found in [Pa,p.169].

8.2 Proposition. *Let N be a noncompact Riemannian n-manifold and suppose that for some $m > n$ and some $c > 0$ the isoperimetric inequality (H^k is the k-dimensional Hausdorff measure)*

$$(8.3) \qquad m(A) \leq cH^{n-1}(\partial A)^{m/(m-1)}$$

holds for all $A \subset\subset N$. Then $\operatorname{cap}(N,C) > 0$ for all continua $C \subset N$.

8.4 Corollary. *Let $f: R^n \longrightarrow N$ be a qr map and suppose that the universal cover $\tilde{N}$ satisfies an inequality (8.3) for all $A \subset\subset \tilde{N}$. Then f is constant.*

For the proof of the corollary we observe that if $\tilde{f}:R^n \longrightarrow \tilde{N}$ is a lift of f, it is constant by the argument in 5.13.

8.5 <u>Examples.</u> If N is the connected sum $S^1 \times S^2 \# S^1 \times S^2$, $\pi_1 N$ is a free group of rank two and then $\tilde{N}$ satisfies (8.3) with $m = \infty$. According to Sullivan's terminology, $\tilde{N}$ is open at infinity. Corollary 8.4 shows that every qr map $f:R^3 \longrightarrow N$ is constant. Alternatively, we get this also by using 7.1 since $\tilde{N}$ is quasiconformally equivalent to $R^3 \setminus E$ where E is a Cantor set.

On the other hand, if $N = S^1 \times S^2$, then $\tilde{N}$ is quasiconformally equivalent to $R^3 \setminus \{0\}$ and the Zorič map $R^3 \longrightarrow R^3 \setminus \{0\}$ with the projection $\tilde{N} \longrightarrow N$ gives a qr map $R^3 \longrightarrow N$.

In the first example $\pi_1 N$ has exponential growth in the word metric; for the definition, see [Gr2,p.68]. An interesting example of a compact 3-manifold N with $\pi_1 N$ having polynomial growth is obtained as $N = G/H$ where $G = \tilde{N}$ is the Heisenberg group which is the Lie group of upper triangular matrices of the form

$$\begin{pmatrix} 1 & x & y \\ 0 & 1 & z \\ 0 & 0 & 1 \end{pmatrix}, \quad x,y,z \in \mathbb{R},$$

and $\pi_1 N = H$ is the subgroup of integer entries. If we fix a left invariant Riemannian metric on G, then G satisfies (8.3) with $m = 4$ (see [Pa]), and hence every qr map of

R^3 into N is constant.

Next let us look a little at some simply connected N. If $N = S^2 \times S^2$, we get a nontrivial qr map $f: R^4 \longrightarrow N$ as follows. There exists a 2 to 1 branched cover φ of the 2-torus T^2 onto S^2. We take f to be the projection $R^4 \longrightarrow T^4 = T^2 \times T^2$ followed by $\varphi \times \varphi$. On the other hand, the case $N = S^2 \times S^2 \# S^2 \times S^2$ is an open problem. Also the case $N = S^2 \times S^2 \setminus \{a\}$ is open.

In the Picard type theorem 7.1 the essential property of $R^n \setminus \{a_1, \ldots, a_q\}$ is that the codimension one homology has a sufficiently large number of generators. In fact, the proof can probably be modified so that the claim of 7.1 holds with some $q(n,K)$ for any Riemannian metric on $R^n \setminus \{a_1, \ldots, a_q\}$. It is natural to ask whether other codimension homology has some effect on the mapping problem. We shall exhibit one example of this type where also the fundamental group plays a role. In fact, it can be shown that there exists no non-constant qr map of R^4 into $N = T^4 \# S^2 \times S^2$. In the proof both the nontriviality of $\pi_1 T^4$ and $H_2(S^2 \times S^2)$ are used. After some preparation the argument is similar to the one in the proof of 7.1. Although $\pi_1(S^1 \times S^3)$ is also nontrivial, the same argument does not apply to $S^1 \times S^3 \# S^2 \times S^2$ and this case is again open.

Some other examples and remarks are presented in [Gr2, Chapter VI] and [Gr1].

REFERENCES

[A] Ahlfors, L.V.: Zur Theorie der Überlagerungsflächen. Acta Math. 65 (1935), 157 - 194.

[BI1] Bojarski, B. and Iwaniec, T.: Another approach to Liouville theorem. Math. Nachr. 107 (1982), 253 - 262.

[BI2] ———— Analytical foundations of the theory of quasiconformal mappings in R^n. Ann. Acad. Sci. Fenn. Ser. AI Math. 8(1983), 257-324.

[F] Federer, H.: Geometric Measure Theory. Springer-Verlag, 1969.

[Ge1] Gehring, F.W.: Symmetrization of rings in space. Trans. Amer. Math. Soc. 101 (1961), 499 - 519.

[Ge2] ———— Rings and quasiconformal mappings in space. Trans. Amer. Math. Soc. 102 (1962), 353 - 593.

[GLM] Granlund, S., Lindqvist, P., and Martio, O.: Conformally invariant variational integrals. Trans. Amer. Math. Soc. 277 (1983), 43 - 73.

[Gr1] Gromov, M.: Hyperbolic manifolds, groups and actions. Ann. of Math. Studies Vol. 97, Princeton University Press, Princeton (1981), 183 - 213.

[Gr2] ———— Structures métriques pour les variétés riemanniennes. Notes de cours rédigées par J. Lafontaine et P. Pansu, CEDIC-Fernand-Nathan, Paris, 1981.

[MaR] Marden, A. and Rickman, S.: Holomorphic mappings of bounded distortion. Proc. Amer. Math. Soc. 46 (1974), 226 - 228.

[MRV1] Martio, O., Rickman, S., and Väisälä J.: Definitions for quasiregular mappings. Ann. Acad. Sci. Fenn. Ser. AI Math. 448 (1969), 1 - 40.

[MRV2] ———— Distortion and singularities of quasiregular mappings. Ann. Acad. Sci. Fenn. Ser. AI Math. 465 (1970), 1 - 13.

[MRV3] ———— Topological and metric properties of quasiregular mappings. Ann. Acad. Sci. Fenn. Ser. AI Math. 488 (1971), 1 - 31.

[MS] Martio, O. and Srebro, U.: On the existence of automorphic quasimeromorphic mappings in R^n. Ann. Acad. Sci. Fenn. Ser. AI Math. 3 (1977), 123 - 130.

[MR] Mattila, P. and Rickman, S.: Averages of the counting function of a quasiregular mappings. Acta Math. 143 (1979), 273 - 305.

[Pa] Pansu, P.: An isoperimetric inequality on the Heisenberg group. Proceedings of "Differential Geometry on Homogeneous Spaces", Torino, 1983, 159 - 174.

[Pe1] Pesonen, M. : A path family approach to Ahlfors's value distribution theory. Ann. Acad. Sci. Fenn. Ser. AI Math. Dissertationes 39 (1982), 1 - 32.

[Pe2] ———— Simplified proofs of some basic theorems for quasiregular mappings. Ann. Acad. Sci. Fenn. Ser. AI Math. 8 (1983), 247 - 250.

[Po] Poleckii, E.A.: The modulus method for non-homeomorphic quasiconformal mappings (Russian). Math. Sb. 83 (1970), 261 - 272.

[Re1] Reshetnyak, J.G.: Sapce mappings with bounded distortion (Russian). Sibirsk. Mat. Ž. 7 (1967), 629 - 658.

[Re2] ———— The Liouville theorem with minimal regularity conditions (Russian). Sibirsk. Mat. Ž 8 (1967), 835 - 840.

[Re3] ———— On the condition of the boundedness of index for mappings with bounded distortion (Russian). Sibirsk. Math. Z. 9 (1968), 368-374.

[Re4] ———— On extremal properties of mappings with bounded distortion (Russian). Sibirsk. Mat. Ž 10 (1969), 1300 - 1310.

[Re5] ———— Spatial Mappings with Bounded Distortion (Russian). Izdatel'stwo "Nauka", Sibirsk. Otdelenie, Novosibirsk, 1982.

[Ri1] Rickman, S.: Path lifting for discrete open mappings. Duke Math. J. 40 (1973), 187 - 191.

[Ri2] ———— On the value distribution of quasimeromorphic maps. Ann. Acad. Sci. Fenn. Ser. AI Math. 2 (1976), 447 - 466.

[Ri3] ———— On the number of omitted values of entire quasiregular mappings. J. Analyse Math. 37 (1980), 100 - 117.

[Ri4] ———— A defect relation for quasimeromorphic mappings. Ann. of Math. 114 (1981), 165 - 191.

[Ri5] ———— Value distribution of quasiregular mappings. Proceedings of the Nordic Summer School in Value

Distribution, Joensuu, 1981. Lecture Notes in Mathematics 981 (1983), 220 - 245, Springer-Verlag.

[Ri6] ———— Quasiregular mappings and metrics on the n-sphere with punctures. Comment. Math. Helvetici 59 (1984), 136 - 148.

[Ri7] ———— The analogue of Picard's theorem for quasiregular mappings in dimension three. Acta. Math. 154 (1985), 195 - 242.

[Ri8] ———— Quasiregular Maps, in preparation.

[V1] Väisälä, J.: Lectures on n-Dimensional Quasiconformal Mappings. Lecture Notes in Mathematics 229, Springer-Verlag, 1971.

[V2] ———— Modulus and capacity inequalities for quasiregular mappings. Ann. Acad. Sci. Fenn. Ser. AI Math. 509 (1972), 1 - 14.

[Zi] Ziemer, W.: Extremal length and p-capacity. Michigan Math. J. 16 (1969), 43 - 51.

[Zo] Zorič, V.A.: The theorem of M.A. Lavrentiev on quasiconformal mappings in space (Russian). Mat. Sb. 74 (1967), 417 - 433.

Conformal and Isometric Immersions of Conformally Flat Riemannian Manifolds into Spheres and Euclidean Spaces

Hans-Bert Rademacher

Contents

Introduction

We consider n-dimensional compact Riemannian manifolds ($n \geqq 3$) which are conformally flat if $n \geqq 4$ and give as well global topological (if $n \geqq 4$) and metric obstructions for the existence of a conformal immersion into the N-dimensional sphere S^N with $N \leqq 2n-2$ (which are due to [Moore 2] and [Moore 3]) as local metric obstructions for the existence of an isometric immersion into S^N or Euclidean space E^N. Then we apply these results to examples of conformally flat manifolds as space forms, products of space forms with opposite curvature and warped products of S^1 and a nonspherical space form.

In [Moore 2] it is proved that a conformally flat compact n-dimensional submanifold M^n of E^N with codimension $N-n \leqq \frac{n}{2}-1$ has vanishing homology groups $H_k(M^n;G) = 0$ for

$N-n<k<2n-N$ for every coefficient module G. We prove this result using the method of [Moore 3] where it is shown that a conformally flat submanifold of S^N can be lifted locally to a flat submanifold of the future-half of the light cone in the (N+2)-dimensional Minkowski space $E^{1,N+1}$.

In Chapter 1 we study the linear algebra of the second fundamental form of a flat submanifold of a scalar product space, in Chapter 2 we study flat submanifolds of the light cone in Minkowski space. We prove in Chapter 3 that a compact n-dimensional Euclidean space form does not admit a conformal immersion in E^{2n-2} or S^{2n-2} ([Moore 3], Theorem 1) and then the above cited theorem. Up to Chapter 3 we do not need the vanishing of the Weyl conformal tensor. In Chapter 4 we prove that a compact n-dimensional Riemannian manifold which is conformally flat if $n \geqq 4$ and which admits a conformal immersion in S^N with $N \leqq 2n-2$ has a point p at which the positive index (i.e. the number of positive eigenvalues) of the Schouten tensor is at least 1 for $N = 2n-2$ ([Moore 3], Theorem 2) or $2n-N$ for $N \leqq 2n-3$. In Theorem 4.4 we show that the Schouten tensor of a n-dimensional submanifold of S^N (resp. E^N) which is conformally flat if $n \geqq 4$ has positive index (resp. positive index plus nullity) at least 1 for $N = 2n-2$ and $2n-N$ for $N \leqq 2n-3$. Then we give necessary and sufficient conditions in terms of the Ricci curvature for the Schouten tensor to be negative (semi-)definite. So a compact n-dimensional Riemannian manifold M^n with $n \geqq 3$ which is conformally flat if $n \geqq 4$ and has Ricci curvature $-c^2 \leqq \mathrm{Ric} \leqq -\frac{n-1}{2n-3}c^2$ with a constant $c \in \mathbb{R}$ cannot be conformally immersed in S^{2n-2}. As examples of conformally flat manifolds Moore studies space forms, we also determine the positive index of the Schouten tensor of Riemannian products of space forms with opposite curvature and warped products of S^1 with a nonspherical spaceform to get obstructions for the existence of conformal or isometric immersions of these manifolds in S^N or E^N (Theorems 4.10, 4.11).

1. Flat symmetric bilinear forms (cf.[Moore 2] ch.2)

If we have an isometric immersion $f : M^n \longrightarrow W^N$ of a flat n-dimensional Riemannian manifold (M^n,g) into a N-dimensional vector space W^N with scalar product $<,>$ then the second fundamental form $\beta : T_pM \times T_pM \longrightarrow N_pM$ at any point $p \in M$ is a symmetric bilinear form on the tangent space T_pM at p with values in the normal space N_pM at p which satisfies the Gauß equation

$$<\beta(x,z), \beta(y,w)> - <\beta(x,w), \beta(y,z)> = 0$$

for all $x,y,z,w \in T_pM$ (cf. [O'Neill] p.100). This leads to the following

1.1 Definition. Let V vesp. W be a n-resp. m-dimensional real vector space. A symmetric bilinear form

$$\beta : V \times V \longrightarrow W$$

with values in W is _flat_ with respect to a non-degenerate real valued symmetric bilinear form (also called _scalar product_)

$$<,>: W \times W \longrightarrow \mathbb{R}$$

iff for all $x,y,z,w \in V$:

$$<\beta(x,z), \beta(y,w)> - <\beta(x,w), \beta(y,z)> = 0 \quad .$$

A scalar product $<,>$ is _Lorentzian_ if there is a basis $e_1,\dots,e_m$ of W with $<e_1,e_1> = -1$, $<e_i,e_i> = 1$ for $i \geq 2$ and $<e_i,e_j> = 0$ if $i \neq j$.

For a symmetric bilinear form $\beta : V \times V \longrightarrow W$ and a vector $x \in V$ we have the linear map $\beta(x) : V \longrightarrow W$, $\beta(x)(y) = \beta(x,y)$ and we set $N(\beta,x) = \ker \beta(x)$. Let $q = \max\{\mathrm{rk}\beta(x) ; x \in V\}$ then $x \in V$ is _regular_ if $\mathrm{rk}\beta(x) = q$. $N(\beta) = \{x \in V ; \beta(x) = 0\}$ is the _nullity space_ of β .

<u>1.2 Lemma</u>. Let $\beta : V \times V \longrightarrow W$ be a flat symmetric bilinear form w.r.t. a scalar product $\langle , \rangle$ on W.

a) If $x \in V$ is regular then for all $n \in N(\beta,x)$:

$$\operatorname{Im}\beta(n) \perp \operatorname{Im}\beta(x) \quad \text{and} \quad \operatorname{Im}\beta(n) \subset \operatorname{Im}\beta(x)$$

b) If $\langle , \rangle$ is positive definite then

$$\dim N(\beta) \geqq \dim V - \dim W .$$

<u>Proof</u>. Since β is flat we have

$$\langle \beta(n,y),\ \beta(z,x) \rangle = \langle \beta(n,x),\ \beta(z,y) \rangle = 0$$

for all $y,z \in V$ since $n \in N(\beta,x)$.

Now we show $\operatorname{Im}\beta(n) \subset \operatorname{Im}\beta(x)$: For $y \in V$ set $g : \mathbb{R} \longrightarrow \mathbb{R}$, $g(t) = \operatorname{rk}\beta(x+ty)$. We have $g(0) = q$ and we show that there is a $\delta > 0$ with $g(t) = q$ for $|t| < \delta$. If $\beta(x,z_1),\ldots,\beta(x,z_q)$ are linearly independent and if $(\beta(x,z_1),\ldots,\beta(x,z_q),\ e_{q+1},\ldots,e_m)$ is a basis for W then there is a $\delta > 0$ such that for all $t \in \mathbb{R}$ with $|t| < \delta$

$$\det(\beta(x+ty,z_1),\ldots,\beta(x+ty,z_q),\ e_{q+1},\ldots,e_m) \neq 0 .$$

For $t \neq 0$ we have $\beta(x+ty,n) = t\beta(y,n)$ i.e. $\beta(y,n) \in \operatorname{Im}\beta(x+ty)$. Now we show by contradiction that this is also true for $t = 0$. Assume $\beta(n,y) \notin \operatorname{Im}\beta(x)$ then there is a basis $(\beta(x,z_1),\ldots,\beta(x,z_q),\ \beta(n,y),\ e_{q+2},\ldots,e_m)$ of W and since

$$\det(\beta(x,z_1),\ldots,\beta(x,z_q),\ \beta(n,y),\ e_{q+2},\ldots,e_m) \neq 0$$

for small $t \in \mathbb{R}$ we have a contradiction.

Now the proof of b) is easy: For a regular $x \in V$ and $n \in N(\beta,x)$ we have from a) $\operatorname{Im}\beta(n) \perp \operatorname{Im}\beta(x)$ and $\operatorname{Im}\beta(n) \subset \operatorname{Im}\beta(x)$ so $\beta(n) = 0$ since $\langle , \rangle$ is positive definite. So we have $N(\beta,x) \subset N(\beta)$, therefore $N(\beta,x) = N(\beta)$ since clearly $N(\beta) \subset N(\beta,x)$. So $\dim N(\beta) = \dim\ker(\beta(x) : V \longrightarrow W) \geqq \dim W - \dim V$.

■

In the sequel we will use the

1.3 Proposition. Let $(W,\langle,\rangle)$ be a Lorentzian scalar product space and $\beta : V \times V \longrightarrow W$ be a symmetric bilinear form flat w.r.t. $\langle,\rangle$. If $\dim V > \dim W$ and $\beta(x,x) \neq 0$ for all $x \neq 0$ then there is a lightlike vector $e \in W$ (i.e. $\langle e,e\rangle = 0,\ e \neq 0$) and a nonzero symmetric bilinear form $\varphi : V \times V \longrightarrow \mathbb{R}$ with

$$\dim N(\beta - e\varphi) \geqq \dim V - \dim W + 2 \quad .$$

Proof. Let $x \in V$ be regular, since $\dim V > \dim W$ there is a $n \in N(\beta,x)$, $n \neq 0$. From 1.2 a) we have for all $y \in V$ a $z \in V$ with $\beta(n,y) = \beta(x,z)$ and

$$\langle\beta(n,y),\ \beta(x,z)\rangle = \langle\beta(n,y),\ \beta(n,y)\rangle = 0 \quad .$$

So $e_1 = \beta(n,n) \neq 0$ is a lightlike vector and let $\beta(n,n) = \beta(x,z)$

$$\langle\beta(n,y),e_1\rangle = \langle\beta(n,y),\beta(n,n)\rangle =$$
$$= \langle\beta(n,y),\beta(x,z)\rangle = \langle\beta(n,x),\beta(y,z)\rangle = 0$$

so $\beta(n,y) = \alpha(y)e_1$; $\alpha(y) \in \mathbb{R}$ because two orthogonal lightlike vectors are linearly dependent. So we have for all $y,z \in V$

$$\langle\beta(y,z),e_1\rangle = \langle\beta(y,z),\beta(n,n)\rangle =$$
$$= \langle\beta(n,y),\beta(n,z)\rangle = \alpha(y)\alpha(z)\langle e_1,e_1\rangle = 0 \quad .$$

Now extend (e_1) to a basis $(e_1,e_2,\ldots,e_m)$ with $\langle e_1,e_2\rangle = \langle e_i,e_i\rangle = 1$ for $i \geqq 2$ and $\langle e_i,e_j\rangle = 0$ otherwise and write $\beta = \sum_{i=1}^{m} e_i\varphi^i$ with m symmetric bilinear realvalued forms $\varphi^i : V \times V \longrightarrow \mathbb{R}$, since $\langle\beta(y,z),e_1\rangle = 0$ for all $y,z \in V$ $\varphi^2 = 0$ and $\varphi^1(n,n) = 1$ so $\varphi^1 \neq 0$. Set $\tilde{W} = \mathrm{span}(e_3,\ldots,e_m)$, $\tilde{\beta} = \beta - e_1\varphi^1$ then $\tilde{\beta}(V \times V) \subset \tilde{W}$. Since $\langle e_1,e_1\rangle = 0$ $e_1\varphi^1$ is flat, so $\tilde{\beta}$ is flat because $\langle\beta(y,z),e_1\varphi^1(u,w)\rangle = 0$ for all $y,z,u,w \in V$. The restriction $\langle,\rangle \mid \tilde{W} \times \tilde{W}$ is positive definite so from Lemma 1.2 b) we can conclude $(e = e_1, \varphi = \varphi^1)$ $\dim N(\beta - e\varphi) \geqq \dim V - \dim W + 2$.

■

2. Flat submanifolds of the lightcone

Let $E^{1,N+1}$ be the (N+2)-dimensional Minkowski space, i.e. $\mathbb{R}^{N+2}$ endowed with the Lorentzian scalar product

$$\langle x,y\rangle = -x_0y_0 + \sum_{i=1}^{N+1} x_iy_i \quad .$$

$x \in E^{1,N+1}$ is spacelike if $\langle x,x\rangle > 0$, lightlike if $\langle x,x\rangle = 0$ and $x \neq 0$ and timelike if $\langle x,x\rangle < 0$. A lightlike vector $x = (x_0,\ldots,x_{N+1})$ is future-pointing if $x_0 > 0$.
$C^{N+1} = \left\{x \in E^{1,N+1} \; ; \langle x,x\rangle = 0 \; , \; x_0 > 0\right\}$ is the future half of the light cone. The induced metric on C^{N+1} is a semi-positive definite metric, its nullity space is tangential to the lines through the origin 0 . On the sphere $S^N = \left\{x \in C^{N+1} \; ; \; x_0 = 1\right\}$ $\langle , \rangle$ induces the standard metric. Define

$$\rho : \mathbb{R}^+ \times S^N \longrightarrow C^{N+1} \; ; \rho(x_0,(1,x')) = (x_0,x_0x') = x_0(1,x')$$

$$\pi : C^{N+1} \longrightarrow S^N \; ; \pi((x_0,x')) = x'$$

then we have the

2.1 Lemma. Let $f : M^n \longrightarrow S^N$ be an immersion of a n-dimensional manifold into the sphere then the set $\{\tilde{f}^*\langle , \rangle \; ; \; \tilde{f} : M^n \longrightarrow C^{N+1}, \; \pi\tilde{f} = f\}$ of Riemannian metrics on M^n induced by lifts $\tilde{f}$ of f into the future half of the light cone is a class of conformally equivalent Riemannian metrics on M^n .

Proof. Every lift $\tilde{f} : M^n \longrightarrow C^{N+1}$, $\pi\tilde{f} = f$ of $f : M^n \longrightarrow S^N$ can be written as $\tilde{f}(p) = e^{\lambda(p)}(1,f(p))$ with a smooth function $\lambda : M^n \longrightarrow \mathbb{R}$. So given two lifts $\tilde{f}_1, \tilde{f}_2$ of f we have $\tilde{f}_1 = e^{\lambda}\tilde{f}_2$ for a smooth function $\lambda : M^n \longrightarrow \mathbb{R}$. Since $\langle\tilde{f}_2(p),\tilde{f}_2(p)\rangle = 0$ we have $\langle d\tilde{f}_2(p)(x),\tilde{f}_2(p)\rangle = 0$ for all

$x \in T_pM$ so for all $x \in T_pM$ we have from

$$d\tilde{f}_1(p)(x) = e^{\lambda(p)}(d\tilde{f}_2(p)(x) + d\lambda(p)(x)\tilde{f}_2(p))$$

that

$$\begin{aligned} &< d\tilde{f}_1(p)(x), d\tilde{f}_1(p)(x) > = \\ &= e^{2\lambda(p)}\Big[< d\tilde{f}_2(p)(x), d\tilde{f}_2(p)(x) > + 2d\lambda(p)(x) < d\tilde{f}_2(p)(x), \tilde{f}_2(p) > + \\ &+ (d\lambda(p)(x))^2 < \tilde{f}_2(p), \tilde{f}_2(p) > \Big] = e^{2\lambda(p)} < d\tilde{f}_2(p)(x), d\tilde{f}_2(p)(x) > , \end{aligned}$$

so $\tilde{f}_1^* <,> = e^{2\lambda}\tilde{f}_2^* <,>$.

■

We have as

<u>2.2 Corollary</u>. If M^n is a n-dimensional Riemannian manifold with metric g and $f : M^n \longrightarrow S^N$ is a conformal immersion then there is a lift $\tilde{f} : M^n \longrightarrow C^{N+1}$ into the future half of the light cone, which is an isometric immersion.

<u>Proof</u>. We have the trivial lift $f : M^n \longrightarrow S^N \subset C^{N+1}$ and for an arbitrary smooth function $\lambda : M^n \longrightarrow \mathbb{R}$ the lift $\tilde{f}(p) = e^{\lambda(p)}(1, f(p))$ so by Lemma 2.1 we get

$$\tilde{f}^* <,> = e^{2\lambda} f^* <,> = e^{2\lambda} e^{\mu} g$$

with a smooth function $\mu : M^n \longrightarrow \mathbb{R}$ since f is conformal. Choose $\lambda = -2\mu$, then $\tilde{f}$ is an isometric immersion.

■

A Riemannian manifold (M^n, g) is <u>conformally flat</u> if every point of M^n has an open neighborhood which is conformally equivalent to an open subset of Euclidean space E^n . Let M^n be a conformally flat manifold with a conformal immersion

$\tilde{f}' : U \longrightarrow C^{N+1}$ into the future half of the light cone with $\pi\tilde{f}' = f|U$ such that the induced metric $\tilde{f}'^*\langle,\rangle$ is flat.

To simplify notation we study the second fundamental form of a spacelike submanifold M^n of C^{N+1} (i.e. each tangent vector is spacelike). The second fundamental form

$$\beta : T_pM \times T_pM \longrightarrow N_pM$$

at $p \in M$ is a symmetric bilinear form on the tangent space at p with values in the normal space N_pM at p of the submanifold $M \subset E^{1,N+1}$. If $\gamma : (-\varepsilon,\varepsilon) \longrightarrow M$ is a smooth curve with $\gamma(0) = p$ and $x = \gamma'(0)$ then $\beta(x,x)$ is the normal component of $\gamma''(0)$.

<u>2.3 Lemma</u>. Let M^n be a spacelike submanifold of the future half C^{N+1} of the light cone. Then we have for all $x,y \in T_pM$:

$$\langle \beta(x,y), p \rangle = - \langle x,y \rangle$$

(where we regard p as a lightlike vector).

<u>Proof</u>. Take a smooth curve $\gamma : (-\varepsilon,\varepsilon) \longrightarrow M$ with $\gamma(0) = p$, $\gamma'(0) = x$. Since $\langle\gamma(t),\gamma(t)\rangle = 0$ for all t we get $\langle\gamma(t),\gamma'(t)\rangle = 0$ and $\langle\gamma(t),\gamma''(t)\rangle = - \langle\gamma'(t),\gamma'(t)\rangle$. So $\langle\beta(x,x),p\rangle = \langle\gamma''(0),\gamma(0)\rangle = \langle\gamma'(0),\gamma'(0)\rangle = - \langle x,x\rangle$ since $p \in N_pM$ and by polarization we get the claim.

∎

If M is a flat submanifold of C^{N+1} (i.e. the induced metric is flat) then from the Gauß-equation we get that the second fundamental form $\beta : T_pM \times T_pM \longrightarrow N_pM$ at $p \in M$ of the submanifold $M \subset E^{1,N+1}$ is flat w.r.t. $\langle,\rangle$ (cf. 1.1) and we can describe β by the

<u>2.4 Proposition</u>. If M^n is a flat spacelike submanifold of C^{N+1} , $p \in M^n$, $N \leq 2n-2$ then one of the following conclusions holds:

a) There is a future-pointing lightlike normal vector $e \in N_pM$ and a subspace V_p of T_pM with $\dim V_p \geq 2n - N \geq 2$ such that for all $x \in V_p$ and $y \in T_pM$

$$\beta(x,y) = e < x,y >$$

(i.e. there is an orthonormal basis $(e_1,\dots,e_n)$ of T_pM with $\beta(e_i,e_j) = e\delta_{ij}$ for $N - n < i \leq n$ and all j) .

b) There is an orthonormal basis $(\bar{e}_1,\dots,\bar{e}_n)$ of N_pM (i.e. $\dim N_pM = N + 2 - n = \dim T_pM = n$) and exactly one future-pointing timelike vector $\bar{e}_1$ with $< \bar{e}_1,p > = -1$ and an orthonormal basis $(e_1,\dots,e_n)$ of T_pM such that

$$\beta(e_i,e_j) = \overline{e_i}\,\delta_{ij}$$

for all i,j . (So if $N < 2n - 2$ only case a) is possible and for $N = 2n - 2$ both cases can occur).

<u>Proof</u>. a) If $\dim N_pM = N + 2 - n < n = \dim T_pM$ we use Proposition 1.3. Since $< \beta(x,x),p > = - < x,x >$ for all $x \in T_pM$ we have $\beta(x,x) \neq 0$ for all $x \neq 0$. So there is a lightlike vector $e \in N_pM$ and a nonzero symmetric bilinear form $\varphi : T_pM \times T_pM \to \mathbb{R}$ with $\dim N(\beta - e\varphi) \geq 2n - N$. So for all $x \in V_p := N(\beta - e\varphi)$ and $y \in T_pM$ we get $\beta(x,y) = e\varphi(x,y)$ and $<\beta(x,y),p> = \varphi(x,y) < e,p > = - < x,y >$. We can substitute φ and e by $\lambda\varphi$ and $\frac{1}{\lambda}e$ for $\lambda \neq 0$ such that $< e,p > = -1$ i.e. e is future-pointing and $\varphi(x,y) = < x,y >$ so $\beta(x,y) = e < x,y >$.

b) If $\dim N_pM = \dim T_pM$ we use results of [Moore1] : Set $S(\beta) = \{\beta(x,y) ; x,y \in T_pM\} \subset N_pM$. If $S(\beta) \neq N_pM$ then from Corollary 3 of Theorem 1 in [Moore1] and from the above argument we get case a). If $S(\beta) = N_p(M)$ then β is nondegenerate and by Theorem 2b) of [Moore1] we have the existence of an orthonormal basis $(\bar{e}_1,\dots,\bar{e}_n)$ of N_pM and n real forms $\theta_i : T_pM \longrightarrow \mathbb{R}$ $i = 1,\dots,n$ with

$\beta(x,y) = \sum_{i=1}^{n} \overline{e_i}\, \theta_i(x)\, \theta_i(y)$. Since $<\beta(x,y),p> = -<x,y>$ $\theta_1,\dots,\theta_n$ are linearly independent and we can choose $\overline{e_i}$ and θ_i such that $<\overline{e_i},p> = -1$ for all $i = 1,\dots,n$ and exactly one - say $\overline{e_1}$ - is future-pointing timelike. Hence $(\theta_1,\dots,\theta_n)$ is an orthonormal basis for the dual space T_p^*M of T_pM and we have the dual orthonormal basis $(e_1,\dots,e_n)$ for T_pM with $\beta(e_i,e_j) = \overline{e_i}\,\delta_{ij}$.

■

3. Conformal immersions of flat and conformally flat manifolds into the sphere

Using Corollary 2.2 and Proposition 2.4 we can show

3.1 Theorem ([Moore3]). If M^n is a compact flat n-dimensional Riemannian manifold, then there does not exist a conformal immersion $f : M^n \longrightarrow S^{2n-2}$.

Remark. Since S^{2n-2} minus a point can be mapped onto the Euclidean space E^{2n-2} by the sterographic projection which is conformal there also does not exist a conformal immersion $f : M^n \longrightarrow E^{2n-2}$ if M^n is flat and compact.

Proof. Assume $f : M^n \longrightarrow S^N$ is a conformal immersion, $N \leq 2n-2$. Then by Corollaray 2.2 we have a lift $\tilde{f} : M^n \longrightarrow C^{N+1}$ into the future half of the light cone which is an isometric immersion i.e. M^n can be seen as a flat submanifold of C^{N+1} . Let $v \in E^{1,N+1}$ be a future-pointing unit timelike vector (i.e. $v = (v_0,v_1,\dots,v_{N+1})$; $v_0 > 0$, $<v,v> = -1$) and $h_v : M^n \longrightarrow \mathbb{R}$, $h_v(p) = -<\tilde{f}(p),v>$ be the height function in the direction of v . For a critical point p of h_v (then $v \in N_pM$) we get for the Hessian

$$d^2h_v(p)(x,y) = -<\beta(x,y),v>$$

for all $x,y \in T_pM$. Now from 2.4 we have in case a) a basis $(e_1,\ldots,e_n)$ of T_pM with $<\beta(e_i,e_j),v> = <e,v>\delta_{ij}$ for all i with $N-n < i \leqq n$ and all j and $<e,v> < 0$ since e is future-pointing. In case b) we have an orthonormal basis $(e_1,\ldots,e_n)$ of T_pM with $<\beta(e_i,e_j),v> = -<\overline{e_i},v>\delta_{ij}$ for all i,j and $<\overline{e_1},v> < 0$ since $\overline{e_1}$ is future-pointing timelike. So in both cases there is $x \in T_pM$ with $d^2h_v(p)(x,x) > 0$. Since M^n is compact h_v has a maximum $p \in M^n$ where for all $x \in T_pM$ $d^2h_v(p)(x,x) \leqq 0$ which yields the contradiction.

■

We need the following local result.

3.2 Lemma. Let M^n be a conformally flat submanifold of the future half C^{N+1} of the light cone, $p \in M^n$ and β its second fundamental form at p. Then there is an open neighborhood U in M of p and a smooth function $\lambda : U \longrightarrow \mathbb{R}^+$ with $\lambda(p) = 1$, $d\lambda(p) = 0$ such that the submanifold $\lambda U = \{\lambda(q)q\,;\,q \in U\}$ of C^{N+1} is flat. Since $\lambda(p) = 1$, $d\lambda(p) = 0$ we can identify then tangent spaces T_pM and $T_p(\lambda U)$ and the normal spaces N_pM and $N_p(\lambda U)$ at p. Then we have for the second fundamental form β_λ of λU at p:

$$\beta_\lambda = \beta + d^2\lambda(p)p \quad .$$

Proof. Since M is conformally flat there is by Corollary 2.2 an open neighborhood U of p and a smooth function $\lambda : U \longrightarrow \mathbb{R}^+$ such that λU is a flat submanifold of C^{N+1} and such that for $p \in M^n$ we have $d\lambda(p) = 0$ and $\lambda(p) = 1$. So we can identify the tangent and normal spaces of U and λU at p. Let $\gamma : (-\varepsilon,\varepsilon) \longrightarrow U$ be a smooth curve with $\gamma(0) = p$ and $x = \gamma'(0)$ then we have for $\tilde{\gamma} = \lambda\gamma : (-\varepsilon,\varepsilon) \longrightarrow U$

$$\tilde{\gamma}''(0) = \lambda(p)\gamma''(0) + 2d\lambda(p)\gamma'(0) + d^2\lambda(p)(\gamma'(0),\gamma'(0))\gamma(0)$$

and projecting into N_pM gives

$$\beta_\lambda(x) = \beta(x,x) + d^2\lambda(p)(x,x)p \quad .$$

■

Now we can prove the

<u>3.3 Theorem</u> ([Moore2]). If M^n is a conformally flat n-dimensional Riemannian manifold and $f : M^n \longrightarrow S^N$ is a conformal immersion then the singular homology groups $H_k(M;G)$ with arbitrary coefficient module are trivial for all k with

$$N - n < k < 2n - N \quad .$$

(The set $\{ k \in \mathbb{N}; N - n < k < 2n - N \}$ is non-empty iff $N \leqq \frac{3}{2}n - 1$ or iff the codimension $N - n \leqq \frac{1}{2}n - 1$) .

<u>Proof</u>. Let g_1 be the metric on M , we can choose a conformal equivalent metric g such that $f : M^n \longrightarrow S^N \subset C^{N+1}$ is an isometric immersion. Let $v = (\sqrt{2}, v') \in E^{1,N+1}$ be a unit time-like vector with $< v', v'> = 1$ and we can regard v' as unit vector of the Euclidean subspace
$\mathbb{R}^{N+1} = \left\{ (0, x_1, \ldots, x_{N+1}) \in E^{1,N+1} \right\}$

$$h_v : M^n \longrightarrow \mathbb{R}, h_v(p) = - < f(p), v >$$

is the height function of M^n in the direction of v (here we regard $f(p)$ again as lightlike vector in $S^N \subset C^{N+1}$) .
For almost all $v \notin h_v$ is a Morse function i.e. at each critical point $p \in M^n$ the Hessian $d^2h_v(p)$ is non-degenerate. So we can assume h_v to be a Morse function.
$E_1^{N+1} = \left\{ (1, x_1, \ldots, x_{N+1}) \in E^{1,N+1} \right\}$ is an affine euclidean subspace of $E^{1,N+1}$ in which S^N lies. Let $\beta : T_pM \times T_pM \longrightarrow N_p'M$ be the second fundamental form of $f : M^n \longrightarrow E_1^{N+1}$ then by Lemma 3.2 there is a smooth function $\lambda : U \longrightarrow \mathbb{R}^+$ defined on an open neighborhood U of p in M such that for

$\lambda f : U \longrightarrow C^{N+1}$ $(\lambda f)^*\langle,\rangle$ is flat and we have for the second fundamental form $\beta_\lambda : T_pM \times T_pM \longrightarrow N_pM$ of the isometric immersion $\lambda f : U \longrightarrow E^{1,N+1}$ (where N_pM is spanned by N'_pM and p) the equation $\beta_\lambda = \beta + d^2\lambda(p)p$. If $a = (1,0,\ldots,0) \in E^{1,N+1}$ then

$$\beta(x,y) = \beta_\lambda(x,y) + \langle \beta_\lambda(x,y), a \rangle p$$

for all $x,y \in T_pM$ since $\beta(x,y) \in N'_pM$. Since $\{k \in \mathbb{N}; N-n < k < 2n-N\} \neq \emptyset$ iff $N \leqq \frac{3}{2}n - 1$ we can assume

$$\dim N_pM = N - n < \dim T_pM = n \quad .$$

Since β_λ is flat w.r.t. $\langle,\rangle$ we can use Proposition 2.4 case a) and the proof of Theorem 3.1: There is an orthonormal basis $(e_1,\ldots,e_n)$ of T_pM and a lightlike future-pointing vector e with

$$\langle \beta_\lambda(e_i,e_j), v \rangle = \langle e,v \rangle \delta_{ij}$$

for all i,j with $N-n < i \leqq n$ and $\langle e,v \rangle < 0$ since e and v are both future-pointing and e is lightlike and v timelike. So we get for a critical point $p \in M$ of h_v :

$$-\langle d^2h_v(p)(e_i,e_j), v \rangle = \langle \beta(e_i,e_j), v \rangle = \mu\delta_{ij}$$

with $\mu = \langle e,v \rangle + \langle e,a \rangle \langle p,v \rangle$ for all i,j with $N-n < i \leqq n$. Since h_v is a Morse function $\mu \neq 0$ and the index of $d^2h_v(p)$ is $\leqq N-n$ (if $\mu < 0$) or $\geqq 2n-N$ (if $\mu > 0$) . So by the fundamental theorem of Morse theory (see [Milnor]) we get the claim.

∎

A _space form_ is a connected complete Riemannian manifold of constant sectional curvature k , if $k > 0$ it is called _spherical_ space form which then is compact, if $k = 0$ it is called _euclidean_ and for $k < 0$ is is called _hyperbolic_.

Space forms are the simplest examples of conformally flat manifolds. That there is no conformal immersion of a n-dimensional compact Euclidean space form into S^{2n-2} was proved in Theorem 3.1, in 4.2 resp. 4.5 we will see that this also holds for compact hyperbolic space forms. For spherical space form we get (cf. [Moore 2]) as

<u>3.4 Corollary</u>. A n-dimensional spherical space form M^n which posesses a conformal immersion in S^N with $N \leqq \frac{3}{2}n - 1$ is isometric to a standard sphere of the same curvature.

<u>Proof</u>. M^n is a Riemannian quotient S^n_r/Γ where $S^n_r := \{x \in \mathbb{R}^{n+1} ; \|x\| = r\}$ and Γ is a finite subgroup of the isometry group $0(n+1)$ of S^n_r. We assume Γ to be nontrivial so there is a cyclic subgroup $\mathbb{Z}_p$ in Γ of prime order p. Let $\pi : S^n/\mathbb{Z}_p \longrightarrow S^n/\Gamma$ be the Riemannian covering so $\pi f : S^n/\mathbb{Z}_p \longrightarrow S^N$ is also a conformal immersion. Since S^n is $(n-1)$-connected $S^n/\mathbb{Z}_p$ is a classifying space for $\mathbb{Z}_p$ up to dimension n so $H_k(S^n/\mathbb{Z}_p ; \mathbb{Z}_p) \cong \mathbb{Z}_p$ for $0 \leqq k < n$. By 3.3 πf cannot be a conformal immersion.

■

Since the connected sum $M_1 \# M_2$ of two conformally flat manifolds M_1 and M_2 has a conformally flat metric we get as

<u>3.5 Corollary</u>. Let M^n_+ be a n-dimensional spherical space form which is not simply-connected and let M^n_1 be any compact connected conformally flat n-dimensional manifold and $M^n = M^n_+ \# M^n_1$ be the connected sum with a conformally flat metric then there is no conformal immersion $f : M^n \longrightarrow S^N$ into the sphere with $N \leqq \frac{3}{2}n - 1$.

<u>Proof</u>. Since for the homology groups of a connected sum we have

$$H_k(M^n ; \mathbb{Z}_p) \cong H_k(M^n_+ ; \mathbb{Z}_p) \oplus H_k(M^n_1 ; \mathbb{Z}_p)$$

for $0 < k < n-1$ we get from the proof of 3.4 that $H_k(M^n; \mathbb{Z}_p)$ is non-trivial for a prime number p and all k with $0 < k < n-1$ which contradicts Theorem 3.3.

■

Remark. If $N = n+1$ in Theorem 3.3 we get that a conformally flat hypersurface of E^{n+1} or S^{n+1} can only have non vanishing homology groups in dimension $0, 1, n-1, n$ and that at each point there is a principal curvature (i.e. eigenvalue of β) which has multiplicity $(n-1)$ (cf. [Lafontaine] D.3).

4. Local and global metric obstructions for n-dimensional Riemannian manifolds with vanishing Weyl conformal tensor to have a conformal or isometric immersion into S^{2n-2} or E^{2n-2}

Let V be finite-dimensional real vector space $h, k : V \times V \longrightarrow \mathbb{R}$ two symmetric bilinear forms, then we have the Kulkarni-Nomizu product $h \cdot k$

$$h \cdot k(x,y,z,w) = h(x,z)k(y,w) + h(y,w)k(x,z) - h(x,w)k(y,z) - h(y,z)k(x,w)$$

$h \cdot k$ is then a 4-covariant curvature tensor (cf. [Lafontaine] B).

If R is the curvature tensor of a Riemannian metric g, Ric the Ricci tensor and s the scalar curvature then the Schouten tensor h is defined by

$$h = \frac{1}{n-2}\left[\mathrm{Ric} - \frac{s}{2(n-1)} g \right] \qquad (n \geq 3)$$

and $W = R - g \cdot h$ is the Weyl conformal tensor. For $n = 3$ $W = 0$ for all metrics, for $n \geq 4$ $W = 0$ iff M^n is conformally flat (cf. [Lafontaine] C).

Now if V and W are finite-dimensional real vector spaces and $h,k : V \times V \longrightarrow W$ are symmetric bilinear forms with values in W and if $<,>$ is a scalar product of W we can define similarly to the Kulkarni-Nomizu product the product $*$:

$$h * k(x,y,z,w) = < h(x,z),k(y,w)> + < h(y,w),k(x,z)> -$$
$$- < h(x,w),k(y,z)> - < h(y,z),k(x,w)>$$

which again is a 4-covariant curvature tensor. So a symmetric bilinear form $\beta : V \times V \longrightarrow W$ is flat w.r.t. $<,>$ iff $\beta * \beta = 0$. (cf. Definition 1.1). For two symmetric bilinear forms $\Psi : V \times V \longrightarrow \mathbb{R}$, $k : V \times V \longrightarrow W$ and a vector $p \in W$ we get:

$$(1) \qquad k * (\Psi p) = \Psi \cdot < k,p > \quad .$$

With this product we can write the Gauß equation of an isometric immersion $f : M^n \longrightarrow W$ into a vector space with scalar product $<,>$ and second fundamental form $\beta : T_pM \times T_pM \longrightarrow N_pM$ as

$$2R = \beta * \beta \quad .$$

Now we get the

<u>4.1 Proposition</u>. Let (M^n,g) be a n-dimensional Riemannian manifold $(n \geq 3)$ and h its Schouten tensor.

a) If $f : M^n \longrightarrow C^{N+1}$ is an isometric immersion into the future half of the light cone (or especially $f : M^n \longrightarrow S^N \subset C^{N+1}$) and $\beta : T_pM \times T_pM \longrightarrow N_pM$ the second fundamental form of $f : M^n \longrightarrow C^{N+1} \subset E^{1,N+1}$ at $p \in M^n$ then we get for the Weyl conformal tensor W at p :

$$2W = (\beta + hf(p)) * (\beta + hf(p))$$

b) $$E^N := \left\{(0,0,x_2,\ldots,x_{N+1}) \in E^{1,N+1}\right\}$$

is an Euclidean subspace of the Minkowski space introduced in Chapter 2 and set

$$q = (1/\sqrt{2}, 1/\sqrt{2}, 0, \ldots, 0) \quad r = (-1/\sqrt{2}, 1/\sqrt{2}, 0, \ldots, 0) \quad .$$

If $f : M^n \longrightarrow E^N$ is an isometric immersion into Euclidean space and $\beta : T_pM \times T_pM \longrightarrow N_pM$ its second fundamental form at $p \in M$ then we get for the Weyl conformal tensor W at p:

$$2W = (\beta + gq - hr) * (\beta + gq - hr) \quad .$$

Proof.

a) $(\beta + hf(p)) * (\beta + hf(p))$

$= \beta * \beta + 2(hf(p)) * \beta$, since $< f(p), f(p) > = 0$

$= 2R - 2h \cdot g$, since $< f(p), \beta > = -g$ (cf. 2.3)

$= 2W$

b) $< q,q > = < r,r > = 0$, $q,r \perp E^N$, $< q,r > = 1$

so

$$(\beta + gq - hr) * (\beta + gq - hr) = \beta * \beta - 2g \cdot h = 2R - 2g \cdot h = 2W \quad .$$

∎

Now we can prove the

4.2 Theorem (cf. [Moore3]) If (M^n, g) is a compact n-dimensional Riemannian manifold $(n \geq 3)$ with Schouten tensor h which is conformally flat if $n \geq 4$ and if $f : M^n \longrightarrow S^N$, $N \leq 2n - 2$ is a conformal immersion then there is a tangent vector $x \in TM$ with $h(x,x) > 0$. If $N < 2n - 2$ then there is a point $p \in M$ where the positive index (i.e. the number of positive eigenvalues) of h is $\geq 2n - N$.

Proof. By Corollary 2.2 there is a lift $\tilde{f} : M^n \longrightarrow C^{N+1}$ into the future-half of the light cone which is an isometric immersion. Let $v \in E^{1,N+1}$ be a future-pointing timelike unit vector and $h_v : M^n \longrightarrow \mathbb{R}$; $h_v(p) = -< \tilde{f}(p), v >$ the height function in the direction of v . Since M^n is compact there is a maximum $p \in M^n$ of h_v . For the second fundamental form

$\beta : T_pM \times T_pM \longrightarrow N_pM$ of $\tilde{f} : M^n \longrightarrow E^{1,N+1}$ at $p \in M$ we have

$$d^2h_v(p)(x,y) = -<\beta(x,y),v>$$

for all $x \in T_pM$ and so $<\beta(x,x),v> \geqq 0$ for all $x \in T_pM$ since p is a maximum of h_v. Since by 4.1 a)

$$\beta + h\tilde{f}(p) : T_pM \times T_pM \longrightarrow N_pM$$

is flat ($\tilde{f}(p) \in N_pM$ since $\tilde{f}(p)$ is lightlike cf. the proof of 2.3) because $W = 0$ there is by Proposition 2.4 a vector subspace V_p of T_pM with $\dim V_p = 1$ if $N = 2n-2$ and $\dim V_p \geqq 2n-N$ if $N < 2n-2$ and a future-pointing lightlike or timelike vector e such that

$$e<x,x> = \beta(x,x) + h(x,x)\tilde{f}(p) \quad .$$

So we get

$$0 > <e,v><x,x> = <\beta(x,x),v> + h(x,x) <\tilde{f}(p),v>$$

from which we can conclude $h(x,x) > 0$ for all $x \in V_p$ since $<\tilde{f}(p),v> < 0$ and $<\beta(x,x),v> \geqq 0$.

■

<u>4.3 Remark</u>. In Proposition 3.2 we considered a conformally flat spacelike submanifold M of the sphere S^N and for any $p \in M$ a smooth function $\lambda : U \longrightarrow \mathbb{R}^+$ defined in an open neighborhood of p in M such that λU is a flat submanifold of the future half of the light cone and $\lambda(p) = 1$ $d\lambda(p) = 0$. Then the second fundamental form β_λ of λU considered in the Minkowski space $E^{1,N+1}$ is flat w.r.t. the scalar product $<,>$ of $E^{1,N+1}$ and we have the equation

$$\beta_\lambda = \beta + d^2\lambda(p)p \quad .$$

Then $\beta_\lambda * \beta_\lambda = 0$ yields for the curvature tensor of U at $p : R = d^2\lambda(p) \cdot g$. Since $R = g \cdot h$ and h is uniquely determined by this equation ([Lafontaine],B) we get

$$h = d^2\lambda(p) \quad .$$

Using 4.1 we can also prove the following local result.

4.4 Theorem. Let (M^n,g) be a n-dimensional Riemannian manifold $(n \geq 3)$ which is conformally flat if $n \geq 4$ and let h be its Schouten tensor.

a) If there is an isometric immersion

$$f : M^n \longrightarrow S^N \ , \quad N \leq 2n-2 \quad ,$$

then we have for all points $p \in M^n$ a tangent vector $x \in T_pM$ with $h(x,x) > 0$. If $N < 2n-2$ then for all points the positive index of h is $\geq 2n-N$.

b) If there is an isometric immersion

$$f : M^n \longrightarrow E^N \ , \quad N \leq 2n-2$$

into Euclidean space then we have for all points $p \in M^n$ a tangent vector $x \in T_pM$, $x \neq 0$ with $h(x,x) \geq 0$. If $N < 2n-2$ then the sum of the nullity and the positive index of h is $\geq 2n-N$ (the nullity is the dimension of the nullity space $N(h)$) .

Proof. a) Let $\beta : T_pM \times T_pM \longrightarrow N'_pM$ be the second fundamental form of $f : M^n \longrightarrow E_1^{N+1}$ with $E_1^{N+1} = \left\{(1,x_1,\ldots,x_{N+1}) \in E^{1,N+1}\right\}$ so $S^N = C^{N+1} \cap E_1^{N+1}$. Set

$$\tilde{\beta} = \beta + hf(p) : T_pM \times T_pM \longrightarrow N_pM$$

where N_pM is the normal space of $f : M^n \longrightarrow E^{1,N+1}$, so N'_pM is a spacelike subspace of N_pM of codimension 1, N_pM

is spanned by $N'_p M$ and $f(p)$. Now

$$<\tilde{\beta}(x,x), f(p)> = <\beta(x,x), f(p)> = -<x,x>$$

(cf. 2.3) so $\tilde{\beta}(x,x) \neq 0$ for all $x \neq 0$. $\tilde{\beta}$ is flat w.r.t. $<,>$ as was shown in 4.1 since now $W = 0$. So by Proposition 2.4 and the proof of Theorem 3.1 there is a subspace V_p of T_pM with $\dim V_p = 1$ if $N = 2n-2$ and $\dim V_p \geqq 2n-N$ if $N < 2n-2$ and a future-pointing timelike or lightlike vector e such that for all $x \in V_p$ we have $\tilde{\beta}(x,x) = e<x,x>$. So for $x \in V_p$ we have $<\tilde{\beta}(x,x), \tilde{\beta}(x,x)> \leqq 0$ and with $\tilde{\beta} = \beta + hf(p)$ we get $0 \geqq <\tilde{\beta}(x,x), \tilde{\beta}(x,x)> = <\beta(x,x), \beta(x,x)> - 2h(x,x)<x,x>$ since $<\beta(x,x), f(p)> = -<x,x>$ and so for $x \neq 0$ $\beta(x,x) \neq 0$ and since $\beta(x,x)$ lies in the spacelike subspace N'_pM we have $h(x,x) > 0$ for all $x \in V_p$, $x \neq 0$.

b) $\beta : T_pM \times T_pM \longrightarrow N'_pM$ is the second fundamental form of $f : M^n \longrightarrow E^N$ with $E^{N^p} = \left\{(0,0,x_2,\dots,x_{N+1}) \in E^{1,N+1}\right\}$ and let N_pM be the normal space to T_pM in $E^{1,N+1}$, so N'_pM is a spacelike subspace of N_pM of codimension 2 , N_pM is spanned by N'_pM and $\{q,r\}$ with $q = (1/\sqrt{2}, 1/\sqrt{2}, 0, \dots, 0)$ $r = (-1/\sqrt{2}, 1/\sqrt{2}, 0, \dots, 0)$. We set

$$\tilde{\beta} = \beta + gq - hr$$

as in 4.1 b), then $\tilde{\beta}$ is flat since $W = 0$ and $<\tilde{\beta}(x,x), r> = g$, so $\tilde{\beta}(x,x) \neq 0$ for $x \neq 0$. As above there is a subspace V_p of T_pM with $\dim V_p \geqq 1$ if $N = 2n-2$ and $\dim V_p \geqq 2n-N$ if $N < 2n-2$ such that $\tilde{\beta}(x,x)$ is timelike or lightlike for all $x \in V_p$ $x \neq 0$. Hence

$$0 \geqq <\tilde{\beta}(x,x), \tilde{\beta}(x,x)> = <\beta(x,x), \beta(x,x)> - h(x,x)<x,x>$$

for all $x \in V_p$ and $h(x,x) \geqq 0$ since $\beta(x,x)$ lies in the spacelike subspace $N'_p M$.

■

Now we give necessary and sufficient conditions for the Schouten tensor to be negative (semi)-definite at a point in terms of the eigenvalues of the Ricci tensor. If the Schouten tensor is negative (semi-)definite then all eigenvalues of the Ricci tensor are negative (non-positive) and do not exceed to much, more precisely:

<u>4.6 Proposition</u>. Let (M^n,g) be a Riemannian manifold, $p \in M^n$ and $(e_1,\dots,e_n)$ an orthonormal basis of T_pM with

$$\mathrm{Ric}(e_i,e_j) = \lambda_i \delta_{ij} \;, \quad \lambda_1 \leqq \dots \leqq \lambda_n \quad .$$

a) If h is negative definite, then all eigenvalues λ_i are negative and

$$\frac{2n-3}{i-1} \lambda_i < \lambda_{i-1} \leqq \lambda_i$$

b) If h is negative semi-definite, then all eigenvalues λ_i are non-positive and

$$\frac{2n-3}{i-1} \lambda_i \leqq \lambda_{i-1} \leqq \lambda_i \quad .$$

<u>Remark</u>. This inequalities could be sharpened.

<u>Proof</u>. We use the definition of h given at the beginning of this chapter:

$$0 \geqq 2(n-1)(n-2)h(e_n,e_n) = (2n-2)\lambda_n - \sum_{j=1}^{n} \lambda_j \geqq$$

$$\geqq (2n-2)\lambda_n - n\lambda_n = (n-2)\lambda_n \quad .$$

So $\lambda_n \leqq 0$ and

$$0 \geqq 2(n-1)(n-2)h(e_i,e_i) = (2n-3)\lambda_i - \sum_{j \neq i} \lambda_j \geqq$$

$$\geqq (2n-3)\lambda_i - (i-1)\lambda_{i-1} \quad .$$

■

Now we give sufficient conditions for the Schouten tensor to be negative (semi)-definite at a point p :

4.7 Proposition. If (M^n, g) is a Riemannian manifold, $p \in M^n$ and if for all $x \in T_pM$, $x \neq 0$ the Ricci curvature $Ric(x) = \frac{Ric(x,x)}{g(x,x)}$ satisfies:

a) $$-c^2 \leqq Ric(x) \leqq -\delta_n c^2$$

with $c \in \mathbb{R}$; $\delta_n := \frac{n-1}{2n-3}$ then h is negative semi-definite at p.

b) $-c^2 \leqq Ric(x) < -\delta_n c^2$ then h is negative definite at p.

Proof. Take an orthonormal basis $(e_1, \dots, e_n)$ of T_pM then by the definition of the Schouten tensor we have:

$$2(n-1)(n-2)h(e_i, e_i) = (2n-2)Ric(e_i) - \sum_{j=1}^{n} Ric(e_j) =$$

$$= (2n-3)Ric(e_i) - \sum_{j \neq i} Ric(e_j) \leqq \left[-(2n-3)\frac{n-1}{2n-3} + n-1\right] c^2 = 0 .$$

■

So we have from the Theorems 4.3, 4.4 and the Proposition 4.7 the

4.8 Theorem. Let (M^n, g) be a n-dimensional Riemannian manifold $(n \geqq 3)$ which is conformally flat if $n \geqq 4$ and $\delta_n = \frac{n-1}{2n-3}$, $c \in \mathbb{R}$

a) If M^n is compact and at all points $p \in M$ we have for the Ricci curvature for all $x \in T_pM$

$$-c^2 \leqq Ric(x) \leqq -\delta_n c^2 \quad ; \quad c \in \mathbb{R}$$

then there is no conformal immersion $f : M^n \longrightarrow S^{2n-2}$.

b) If there is a point $p \in M^n$ with

$$-c^2 \leqq Ric(x) \leqq -\delta_n c^2 \quad (\text{or} \quad -c^2 \leqq Ric(x) < -\delta_n c^2)$$

for all $x \in T_pM$ then there is no isometric immersion $f : M^n \longrightarrow S^{2n-2}$ (or $f : M^n \longrightarrow E^{2n-2}$).

<u>4.9 Remarks</u>. a) So obviously a compact n-dimensional hyperbolic or Euclidean space form cannot be conformally immersed in S^{2n-2} . Even locally an n-dimensional hyperbolic space form cannot be isometrically immersed in E^{2n-2} and an Euclidean or hyperbolic n-dimensional space form cannot be locally isometrically immersed in S^{2n-2} (this was shown by [Cartan]).

b) For $n = 3$ there is in Theorem 4.8 no restriction on the metric besides the pinching of the Ricci curvature with $\delta_3 = \frac{2}{3}$. Since the sequence

$$\delta_n = \left(2 - \frac{1}{n-1}\right)^{-1} \in \left(\frac{2}{3}, \frac{1}{2}\right)$$

is monotone decreasing one could choose $\delta_n = \frac{2}{3}$ in Theorem 4.8 to get a pinching-condition which does not depend on the dimension. Now we will give obstructions for Riemannian products $(M_1^{n_1}, g_1) \times (M_2^{n_2}, g_2)$ with vanishing Weyl curvature to have conformal or isometric immersions in the sphere or Euclidean space of low codimension. The Riemannian product $(M_1^{n_1}, g_1) \times (M_2^{n_2}, g_2)$ is conformally flat if and only if $n_1 = 1$ and $M_2^{n_2}$ is a space form or if $M_1^{n_1}$ and $M_2^{n_2}$ are both space forms of dimension at least two with opposite curvatures. ([Lafontaine]D2). Then we get the following

<u>4.10 Theorem</u>. Let $n \geqq 4$, $n - m \geqq 2$, $n, m \in \mathbb{N}$ and let (M_1^m, g_1) be S^1 for $m = 1$ and for $m \geqq 2$ be a compact m-dimensional space form with positive curvature k_1 and let (M_2^{n-m}, g_2) be a compact (n-m)-dimensional space form with negative curvature k_2 and $k_2 = -k_1$ if $m \geqq 2$. Then there is no conformal immersion

$$f : (M^n,g) := (M_1^m,g_1) \times (M_2^{n-m},g_2) \longrightarrow S^N$$

with $N < 2n - \max\{m,2\}$ and there is even locally no isometric immersion of M^n into S^N or E^N if $N < 2n - \max\{m,2\}$.

Proof. Let $p = (p_1,p_2) \in M_1^m \times M_2^{n-m}$ and $(e_1,\ldots,e_m)$ and $(e_{m+1},\ldots,e_n)$ be orthonormal basis for $T_{p_1}M_1$ and $T_{p_2}M_2$. Then we have for $1 \leq i \leq m$ for the Schouten tensor of M at p:

$$(2n-2)(n-2)h(e_i,e_i) = 2(n-1)\ \mathrm{Ric}(e_i) - \sum_{i=1}^{n} \mathrm{Ric}(e_j) =$$

$$= 2(n-1)(m-1)k_1 - m(m-1)k_1 - (n-m-1)(n-m)k_2 =$$

$$= (n-1)(n-2)(-k_2) > 0 \qquad \text{for} \quad n \geq 3$$

and for $m+1 \leq i \leq n$

$$(2n-2)(n-2)h(e_i,e_i) =$$

$$= 2(n-1)(n-m-1)k_2 - m(m-1)k_1 - (n-m)(n-m-1)k_2 < 0 \quad .$$

So the positive index of h at p is m and by Theorem 4.2 resp. 4.4 we get for the positive index of h at a point resp. all points since h is non-degenerate the lower bound $2n - N$ if there is a conformal resp. isometric immersion $M^n \longrightarrow S^N$, $N \leq 2n-2$, so in our case $m \geq 2n - N$ or $N \geq 2n - m$.

∎

Another example of a conformally flat Riemannian manifold is the warped product $M^n = S^1 \times_f M_k^{n-1}$ where M_k^{n-1} is a $(n-1)$-dimensional space form of curvature k, $f : S^1 \longrightarrow \mathbb{R}^+$ is smooth and the metric g on the differentiable product $S^1 \times M_k^{n-1}$ is defined by $dt^2 + f^2(t)g_k$ where $(t,p) \in S^1 \times M_k^{n-1}$, dt^2 the standard metric on S^1 and g_k

the metric of M_k^{n-1} (cf. [O'Neill] p. 204 and [Lafontaine] D.1).

4.11 Theorem. Let $n \geqq 3$ and M_k^{n-1} be a compact (n-1)-dimensional space form with non-positive curvature k, $f : S^1 \longrightarrow \mathbb{R}^+$ be a smooth function and $M^n = S^1 \times_f M_k^{n-1}$ be the warped product. Then there is no conformal immersion of M^n into S^{2n-3} and even locally there is no isometric immersion of M^n into S^{2n-3} (resp. E^{2n-3} if $k < 0$).

Proof. Let $(t,p) \in S^1 \times M_k^{n-1}$ be a point of M^n and $(e_1,\dots,e_n)$ be an orthonormal basis of T_pM such that $e_1 \in T_tS^1 \subset T_{(t,p)}M^n$ and $(e_2,\dots,e_n)$ is an orthonormal basis of $T_pM_k^{n-1} \subset T_{(t,p)}M^n$.

Using Corollary 7.43 of O'Neill we get

$$\operatorname{Ric}(e_1,e_1) = -(n-1)\frac{f''}{f}$$

and for $2 \leqq i \leqq n$

$$\operatorname{Ric}(e_i,e_i) = \frac{k(n-2)}{f^2} - \frac{f''}{f} - (n-2)\frac{f'^2}{f^2}$$

and $\operatorname{Ric}(e_i,e_j) = 0$ for $i \neq j$. The scalar curvature s is then given by

$$\begin{aligned} s &= \operatorname{Ric}(e_1,e_1) + (n-1)\operatorname{Ric}(e_2,e_2) = \\ &= -2(n-1)\frac{f''}{f} + (n-1)(n-2)\frac{k}{f^2} - (n-1)(n-2)\left(\frac{f'}{f}\right)^2 \end{aligned}$$

and so the Schouten tensor can be computed:

$$h(e_1,e_1) = -\frac{k}{2f^2} + \frac{1}{2}\left(\frac{f'}{f}\right)^2 - \frac{f''}{f}$$

$$h(e_i,e_i) = \frac{k}{2f^2} - \frac{1}{2}\left(\frac{f'}{f}\right)^2 \qquad \text{for } 2 \leqq i \leqq n$$

and $h(e_i,e_j) = 0$ for $i \neq j$. (In the second formula the second derivative of f does not enter).

So for $k \leq 0$ the positive index of h is at most 1 and for $k < 0$ the negative index is at least $(n-1)$, therefore the claim follows from the Theorems 4.2 and 4.4.

■

4.12 Corollary. Let $n \geq 3$ and M_0^{n-1} be a $(n-1)$-dimensional Euclidean space form and $f : S^1 \longrightarrow \mathbb{R}^+$ be a smooth function. Then there is no isometric immersion of the warped product $M^n = S^1 \times_f M_0^{n-1}$ into S^{2n-2} .

Proof. $f : S^1 \longrightarrow \mathbb{R}^+$ posseses a minimum $t \in S^1$ with $f'(t) = 0$ and $f''(t) \geq 0$. Then the formulae for the Schouten tensor h in the proof of 4.11 show that h at a point (t,p), $p \in M_0^{n-1}$ is negative semi-definite.

■

References.

E. Cartan: Sur les variétés de courbure constante d'un espace euclidean ou non-euclidean. Bull. Soc. Math. France 47 (1919) 125-160 and 48 (1920) 132-208.

J. Lafontaine: Conformal geometry from the Riemannian viewpoint. This volume.

J. Milnor: Morse Theory. Ann. of Math. Studies 51 Princeton NJ University Press 1969.

J. Moore 1: Submanifolds of constant positive curvature I. Duke Math. J. 44 (1977) 449-484.

J. Moore 2: Conformally flat submanifolds of Euclidean space. Math. Ann. 225 (1977) 89-97.

J. Moore 3: On conformal immersions of space forms. In: Global differential geometry and global analysis. Proceedings Berlin 1979. Lect. Notes in Math. 838 Springer Berlin/Heidelberg/New York 1981.

B. O'Neill: Semi-Riemannian geometry. With applications to relativity. Academic Press New York/London 1983.

Compact Conformally Flat Hypersurfaces

Ulrich Pinkall

Contents

1. Introduction

Which compact n-dimensional conformally flat manifolds (M^n,g) admit a conformal immersion as a hypersurface in $\mathbb{R}^{n+1}$?

For $n = 2$ this problem was solved by A. Garsia ([5], see also [9]), who proved that every compact Riemann surface (M^n,g) (being always conformally flat, see chapter 1 of these notes) can be conformally immersed into $\mathbb{R}^3$.

For conformally flat 3-manifolds there are some known obstructions to conformal immersibility into $\mathbb{R}^4$ [2], but in general the above question is open in this case.

For compact conformally flat hypersurfaces M^n in $\mathbb{R}^{n+1}$, $n \geq 4$ the problem is solved to a large extent. In particular the possible topological types of M^n are known: Let $S^{n-1} \times_n S^1$ denote the n-dimensional Klein bottle (n stands for an orientation reversing isometry of S^{n-1}).

Theorem 1 (do Carmo, Dajzer, Mercuri [3]): Let (M^n,g) be a compact conformally flat manifold, $n \geq 4$, $f: M^n \to \mathbb{R}^{n+1}$ a conformal immersion. Then for some $k \geq 0$ M^n is diffeomorphic to

(i) $\underbrace{(S^{n-1} \times S^1) \times \ldots \times (S^{n-1} \times S^1)}_{k \text{ copies}}$

if M^n is orientable

(ii) $\underbrace{(S^{n-1} \times S^1) \times \ldots \times (S^{n-1} \times S^1)}_{k \text{ copies}} \times (S^{n-1} \times_n S^1)$

if M^n is nonorientable.

Also the extrinsic geometry of compact conformally flat hypersurfaces in $\mathbb{R}^n$, $n \geq 4$ is well understood ([3], see section 2 below). Here we will determine in addition the intrinsic conformal geometry of such hypersurfaces:

Theorem 2: Every compact conformally flat hypersurface in $\mathbb{R}^{n+1}$, $n \geq 4$ is conformally equivalent to a classical Schottky manifold.

Classical Schottky manifolds are constructed as follows: Start with the standard sphere (S^n, g_{can}) and

(i) closed round balls $B_1, \ldots, B_k$ and $\tilde{B}_1, \ldots, \tilde{B}_k$ which are pairwise disjoint.

(ii) Möbius transformations $f_1, \dots f_k : S^n \to S^n$ such that

$$f_i(\mathring{B}_i) = S^n - \tilde{B}_i .$$

Then the quotient space obtained from $S^n - \bigcup_{i=1}^{k} (\mathring{B}_i \cup \mathring{\tilde{B}}_i)$ by identifying ∂B_i with $\partial \tilde{B}_i$ via f_i carries in a canonical way the structure of a compact conformally flat manifold.

Theorem 2 is stronger than Theorem 1 because of

<u>Theorem 3</u>: For $n \geq 3$ and $k \geq 2$ there are conformally flat metrics g on

$$M^n = \underbrace{(S^{n-1} \times S^1) \times \dots \times (S^{n-1} \times S^1)}_{k \text{ copies}}$$

such that (M^n, g) is not conformally equivalent to a classical Schottky manifold.

It still remains an open problem wether every classical Schottky manifold (M^n, g) admits a conformal immersion into $\mathbb{R}^{n+1}$. To illustrate this problem we will look more closely at the case $k = 1$. Orientable Schottky manifolds M^n with $k = 1$ are diffeomorphic to $S^{n-1} \times S^1$ and can be described as

$$M^n_{\lambda,g} = \mathbb{R}^n - \{0\}_{/\lambda g}$$

where $\lambda > 0$, $g \in SO(n)$. In section 4 we will prove

<u>Theorem 4</u>: There is a number $\lambda_0 > 0$ such that for $\lambda > \lambda_0$ all $M^n_{\lambda,g}$ admit conformal immersions into $\mathbb{R}^{n+1}$.

On the other hand there is some evidence for the conjecture that for g far from the identity and λ small $M^n_{\lambda,g}$ cannot be conformally immersed as a hypersurface in $\mathbb{R}^{n+1}$, but we are not able to prove this.

2. Conformally flat hypersurfaces

Throughout this section (M^n,g), $n \geq 4$ will denote a compact conformally flat manifold, $f: M^n \to \mathbb{R}^{n+1}$ a conformal immersion. Recall (E. Cartan [1], see chapter [3]) that at each point $p \in M^n$ there is a principal curvature λ of multiplicity n or n-1. Let $U \subset M^n$ be the set of non-umbillic points, i.e. the set where λ has multiplicity n-1. For $p \in U$ let $D_p \subset T_pM^n$ denote the eigenspace of the second fundamental form which corresponds to λ. The following facts are well-known. The most crucial part, the completeness of the curvature leaves in (ii), is due to Reckziegel [8].

(i) The distribution $p \to D_p$ on U is integrable and therefore leads to a foliation of U by so-called "curvature leaves".

(ii) All curvature leaves are compact and are mapped by f diffeomorphically onto round (n-1)-spheres in $\mathbb{R}^{n+1}$.

(iii) Along each curvature leaf the hypersurface f is tangent to a fixed round hypersphere (or hyperplane) in $\mathbb{R}^{n+1}$.

(iii) follows from the fact that λ is constant along each curvature leaf, which implies the same for the corresponding focal point $f - \lambda N \in \mathbb{R}^{n+1} \cup \{\infty\}$ $(N: M^n \to S^n \subset \mathbb{R}^{n+1}$ the unit normal vector field satisfying $XN = \lambda Xf$ for all tangent vectors $X \in D)$. (iii) can be rephrased by saying that $f|_U$ is the envelope of a 1-parameter family of hyperspheres, i.e. a "channel hypersurface".

Let now $h: \mathbb{R}^{n+1} \to \mathbb{R}$ be an affine function such that $h \circ f: M^n \to \mathbb{R}$ is a Morse function (such h exists [6]). Then $h \circ f$ has a certain number of minima and maxima, all in $M^n - U$, and some critical points $p_1, \dots, p_l \in U$ of index 1 or n-1. Through each point p_i we have a curvature leaf L (an (n-1)-sphere as we know). Taking into account the possibility that there are several p_i on the same leaf we thus have finitely many curvature leaves $L_1, \dots, L_m$ such that on the complement of $L_1 \cup \dots \cup L_m$ all critical points of $h \circ f$ are minima or maxima.

Each leaf L_i has a neighborhood V_i, diffeomorphic to $S^{n-1} \times [-1,1]$ such that ∂V_i consists of two curvature leaves. Each V_i is a simply connected conformally flat manifold with boundary, hence there is a developing map $\sigma_i : V_i \to S^n$ (see chapter 1). f maps ∂V_i to a totally umbillic submanifold of $\mathbb{R}^{n+1}$, which implies that ∂V_i is totally umbillic in (M^n, g). This means that σ maps each component S of ∂V_i to a round sphere $\sigma(S)$ in S^n. Some collar C of S in V_i is mapped to a one-sided collar $\sigma(S)$ in S^n. (Figure 1).

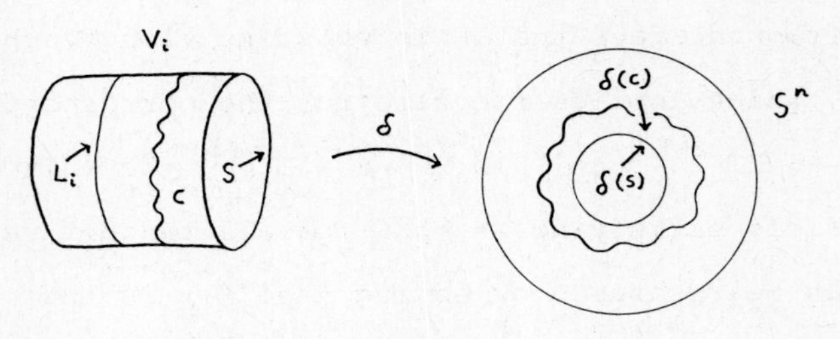

Figure 1

Let B denote the component of $S^n - \sigma(S)$ which is disjoint from $\sigma(C)$. We use σ to glue B to V_i. Proceeding similarly with the second component of ∂V_i we obtain a compact conformally flat manifold $\overline{V}_i$. σ extends to a local homeomorphism $\overline{\sigma}:\overline{V}_i \to S^n$. Since $\overline{V}_i$ is compact, $\overline{\sigma}$ must be a covering map, hence a diffeomorphism. This implies that σ was an embedding. We record this as

Lemma 1: Each V_i is conformally equivalent to

$S^n - (\mathring{B}_1 \cup \mathring{B}_2)$, where B_1, B_2 are disjoint closed round balls in S^n.

Applying the same argument as above to the components $M_1, \ldots, M_q$ of $M^n - \bigcup_{i=1}^{n} V_i$ we obtain compact conformally flat manifolds $\overline{M}_1, \ldots \overline{M}_q$ without boundary. Moreover by fact (iii) recorded at the beginning of this section we can extend $f|_{M_i}$ to a C^1-immersion $\overline{f}_i : \overline{M}_i \to \mathbb{R}^{n+1}$ such that $f|_{\overline{M}_i - M_i}$ is totally umbillic.

Note that the height function $h \circ \overline{f}_i : \overline{M}_i \to \mathbb{R}$ has the following properties:

(i) $h \circ \overline{f}_i$ is a C^1-function which is C^∞ outside $\partial M_i \subset \overline{M}_i$.

(ii) $h \circ \overline{f}_i$ has no critical points on ∂M_i.

(iii) All critical points of $h \circ \overline{f}_i$ are nondegenerate minima or maxima.

Thus we can approximate $h \circ \overline{f}_i$ by a C^∞ Morse function $g : \overline{M}_i \to \mathbb{R}$ all of whose critical points are minima or maxima. By Morse theory then g_i has in fact only one minimum and one maximum and $\overline{M}_i$ is homeomorphic to the sphere S^n. Thus the developing map $\sigma_i : \overline{M}_i \to S^n$ is a diffeomorphism and as in the proof of Lemma 1 we conclude

<u>Lemma 2</u>: Each M_i is conformally equivalent to $S^n - \bigcup_{j=1}^{m_i} \mathring{B}_{ij}$ where $B_{ij} \subset S^n$ are disjoint closed round balls.

Summarizing Lemma 1 and Lemma 2 we can reconstruct the conformally flat manifold (M^n, g) as follows: Start with finitely many copies $S_1^n, \ldots, S_{m+q}^n$ of the standard n-sphere and remove from $S_1^n \amalg \ldots \amalg S_{m+q}^n$ disjoint round balls $B_1, \ldots, B_k$, $\widetilde{B}_1, \ldots, \widetilde{B}_k$. Assuming $B_i \subset S_{a_i}$, $\widetilde{B}_i \subset S_{b_i}$ identify ∂B_i with $\partial \widetilde{B}_i$ via a Möbius transformation $f_i : S_{a_i}^n \to S_{b_i}^n$.

To finish the proof of Theorem 2 we only have to show that (M^n, g) can also be constructed using only a single sphere S^n, cutting round holes and identifying the hole boundaries via Möbius transformations. This can be done using an induction on $m + q$, which is left to the reader.

3. Non-classical Schottky manifolds

There is a slightly more general notion of "Schottky manifolds" than the one given in the introduction: Start with the standard sphere S^n and

(i) smoothly embedded closed n-balls $B_1,\dots,B_k$, $\tilde{B}_1,\dots,\tilde{B}_k$ in S^n which are pairwise disjoint.

(ii) Möbius transformations $f_1,\dots,f_n : S^n \to S^n$ such that

$$f_i(\mathring{B}_i) = S^n - \tilde{B}_i \ .$$

Then as before the quotient space obtained from $S^n - \bigcup_{i=1}^{k} (\mathring{B}_i \cup \mathring{\tilde{B}}_i)$ by identifying ∂B_i with $\partial \tilde{B}_i$ via f_i carries the structure of a conformally flat manifold M^n. Assuming for the sake of simplicity that the f_i are orientation presering it is clear that M^n is diffeomorphic to a connected sum of copies of $S^{n-1} \times S^1$.

If the n-balls $B_1,\dots,B_k$, $\tilde{B}_1,\dots,\tilde{B}_k$ can be chosen as round balls the Schottky manifold is called classical. We will prove Theorem 3 be exhibiting for $k \geq 2$, $n \geq 3$ Schottky manifolds which are not classical.

The construction will use an induction on n. Actually we start the induction with $n = 2$, keeping in mind that a "conformally flat structure" on a two-dimensional surface should be interpreted as a Möbius structure (see chapter 1). For $n = 2$ Theorem 3 was in fact essentially proved by Marden [7].
His proof needs only the following modification:
Marden uses continuously embedded 2-discs $B_1,\dots,B_k$,

$\tilde{B}_1,\dots,\tilde{B}_k$ in his definition of a Schottky surface. However inside M^2 the curves ∂B_i correspond to simple closed curves, which can easily be smoothed. Lifting the smoothed curves back to S^2 we see that every Schottky surface in Marden's sense is also Schottky with our definition.

Suppose now we are given an n-dimensional non-classical Schottky manifold M^n, constructed as above using smooth balls $B_1,\dots,B_k$, $\tilde{B}_1,\dots,\tilde{B}_k \subset S^n$ and Möbius transformations $f_1,\dots,f_k : S^n \to S^n$. Realize S^n as a round hypersphere in S^{n+1}. Then the $f_1,\dots,f_k$ can be uniquely extended to Möbius transformations $g_1,\dots,g_k : S^{n+1} \to S^{n+1}$. We now construct pairwise disjoint smoothly embedded (n+1)-balls $C_1,\dots,C_k$, $\tilde{C}_1,\dots,\tilde{C}_k \subset S^{n+1}$ such that

(i) $\quad C_i \cap S^n = B_i \quad , \quad \tilde{C}_i \cap S^n = \tilde{B}_i$.

(ii) $\quad g_i(\mathring{C}_i) = S^{n+1} - \tilde{C}_i$.

Afterwards we prove that the Schottky manifold defined using $C_1,\dots,C_k$, $\tilde{C}_1,\dots,\tilde{C}_k$ and $g_1,\dots,g_k$ is non-classical. Set

$$C_i = \cup\{\text{round balls } C \subset S^{n+1} \mid \partial C \perp S^n \text{ and } C \cap S^n \subset B_i\}$$

$$\tilde{C}_i = \cup\{\text{round balls } C \subset S^{n+1} \mid \partial C \perp S^n \text{ and } C \cap S^n \subset S^n - \mathring{\tilde{B}}_i\}.$$

Then obviously ∂C_i is C^1 at $\partial C_i \cap S^n$ (in fact ∂C_i is orthogonal to S^n). To see that ∂C_i is C^1 everywhere note that each of the two components of $S^{n+1} - S^n$ can be considered as a Poincaré model of hyperbolic (n+1)-space H^{n+1}. $C_i \cap H^{n+1}$ is then just the convex hull of $B_i \subset S^n = \quad H^{n+1}$. The smoothness of $\partial C_i \cap H^{n+1}$ is then easier to discuss in the projective model of

H^{n+1} and is fact established by the following lemma:

Lemma 3: Given a smoothly embedded n-ball $B \subset S^n$ on the boundary S^n of the standard n-ball $D^{n+1} \subset \mathbb{R}^{n+1}$ let conv(B) denote the convex hull of B in $\mathbb{R}^{n+1}$. Then ∂ conv(B) $\cap D^{n+1}$ is a smooth hypersurface in $\mathbb{R}^{n+1}$.

The proof of Lemma 3 is left to the reader. It is easy to see that $C_1,\dots,C_k$, $\tilde{C}_1,\dots,\tilde{C}_k$ and $g_1,\dots,g_k$ meet all requirements listed in the definition of a Schottky manifold. It remains to show that the corresponding Schottky manifold M^{n+1} is non-classical.

Assume M^{n+1} is classical, i.e. M^{n+1} can also be constucted using round (n+1)-balls $D_1,\dots,D_k$, $\tilde{D}_1,\dots,\tilde{D}_k \subset S^{n+1}$ and Möbius transformations $h_1,\dots,h_k : S^{n+1} \to S^{n+1}$. $M^n \subset M^{n+1}$ is by construction a totally umbillic submanifold of M^{n+1}, whose universal cover $\tilde{M}^n$ is mapped by the developing map $\sigma:\tilde{M}^{n+1} \to S^{n+1}$ to a round n-sphere $S^n \subset S^{n+1}$. S^n is invariant under the holonomy representation of M^{n+1} (see chapter 1 for definitions), in particular under the Möbius transformations $h_1,\dots,h_k$. Hence M^n can be constructed by removing from S^n the round balls $\mathring{D}_1 \cap S^n,\dots,\mathring{D}_k \cap S^n$, $\mathring{\tilde{D}}_1 \cap S^n,\dots,\mathring{\tilde{D}}_k \cap S^n$ and identifying the boundary components of the resulting manifold with boundary in pairs via the Möbius transformations h_i. This contradicts our assumption that M^n is non-classical.

4. Channel tori

In this section we investigate a special class of compact

conformally flat hypersurfaces M^n immerced into $\mathbb{R}^{n+1}$, called channel tori. Our treatment is a modernized version of the one given by Garsia in [4], who was interested in the case $n = 2$. Roughly speaking channel tori are envelopes of one-parameter families of hyperspheres.

The precise meaning of "envelope" is provided by the following setup: Let $M^n \subset S^n \times S^1$ be a compact submanifold and $f : S^n \times S^1 \to \mathbb{R}^{n+1}$ be a smooth map such that the following conditions are satisfied:

(i) $f|_{S^n \times \{t\}}$ is an embedding of $S^n \times \{t\}$ onto a round hypersphere in $\mathbb{R}^{n+1}$ for all $t \in S^1$.

(ii) $M^n \subset S^n \times S^1$ is transversal to the submanifold $S^n \times \{t\}$ for all $t \in S^1$. Moreover for all $(p,t) \in M^n$ we have

$$df(T_{(p,t)}M^n) = df[T_{(p,t)}(S^n \times \{t\})].$$

It is an immediate consequence of (i) and (ii) that $f|_{M^n} : M^n \to \mathbb{R}^{n+1}$ is an immersion. We will show below that M^n with the induced metric from $\mathbb{R}^{n+1}$ is conformally flat, in fact we will construct a conformal diffeomorphism between M^n and $M^n_{\lambda,g}$ (see the introduction) for some $\lambda \in \mathbb{R}$, $g \in SO(n)$. Our main concern will be to obtain some control over λ and the conjugacy class of g in $SO(n)$.

We first investigate the map $\tilde{f} : S^n \times \mathbb{R} \to \mathbb{R}^{n+1}$ induced from f by composing f with the universal covering projection $S^n \times \mathbb{R} = S^n \times S^1 \to S^n \times S^1$. We consider $S^n \times \mathbb{R}$ as an S^n-bundle over $\mathbb{R}$ endowed with the product metric.

Lemma 4: There is a fibre-preserving diffeomorphism

$h: S^n \times \mathbb{R} \to S^n \times \mathbb{R}$ such that the map

$\hat{f} = \tilde{f} \circ h : S^n \times \mathbb{R} \to \mathbb{R}^{n+1}$ has the following properties:

(i) $\hat{f}\big|_{S^n \times t} : S^n \times \{t\} \to \mathbb{R}^{n+1}$ is a conformal embedding onto a round hypersphere for each $t \in \mathbb{R}$.

(ii) $d\hat{f}_{(p,t)}(\frac{\partial}{\partial t}) \perp df[T_{(p,t)}(S^n \times \{t\})]$ for all $(p,t) \in S^n \times \mathbb{R}$.

Proof: We define a time-dependent vector field X^t, $t \in \mathbb{R}$ on S^n as follows: For $p \in S^n$, $t \in \mathbb{R}$ let $X_p^t \in T_p S^n$ be the unique tangent vector (see property (i) at the beginning of this section) satisfying

$$(1) \qquad d\tilde{f}_{(p,t)}(X_p^t, 0) = \begin{cases} \text{orthogonal projection of} & d\tilde{f}_{(p,t)}(\frac{\partial}{\partial t}) \\ \text{onto} & d\tilde{f}[T_{(p,t)}(S^n \times \{t\})]. \end{cases}$$

Here we have used the obvious splitting

$$(2) \qquad T_{(p,t)}(S^n \times \mathbb{R}) = T_p S^n \oplus \mathbb{R}.$$

Moreover let $H_o : S^n \to S^n$ be such that the map $p \mapsto \tilde{f}(H\ (p), 0)$ is conformal and define diffeomorphism $h_t : S^n \to S^n$ by the initial value problem

$$(3) \qquad \begin{aligned} h_o &= H_o \\ \frac{\partial h_t(p)}{\partial t} &= -X_p^t \end{aligned} \quad .$$

It is easy to check that $\hat{f} = \tilde{f} \circ h$ with $h(p,t) = h_t(p)$ satisfies property (ii) of the Lemma. It remains to establish property (i) for this h. Let g_t denote the metric induced on S^n by the embedding $p \to \hat{f}(p,t)$. Then g_o is the standard metric on S^n. It is therefore enough to show that for all t the derivative $\frac{\partial g_t}{\partial t}$ is proportional to g_t. But this is clear, since the variation $\frac{\partial g_t}{\partial t}$ by (ii) is induced from a normal variation of the embedding $p \mapsto \hat{f}(p,t)$. $\frac{\partial g_t}{\partial t}$ is therefore proportional to the second fundamental from of this embedding, which is a multiple of g_t because $p \mapsto \hat{f}(p,t)$ is totally umbillic.

□

What we have accomplished with Lemma 4 is a parametrization of the given one-parameter family of hyperspheres by "orthogonal trajectories" (Figure 2). (i) of Lemma 4 just says that the correspondence between any two spheres in a one-parameter family of hyperspheres induced by the orthogonal trajectories is a conformal diffeomorphism.

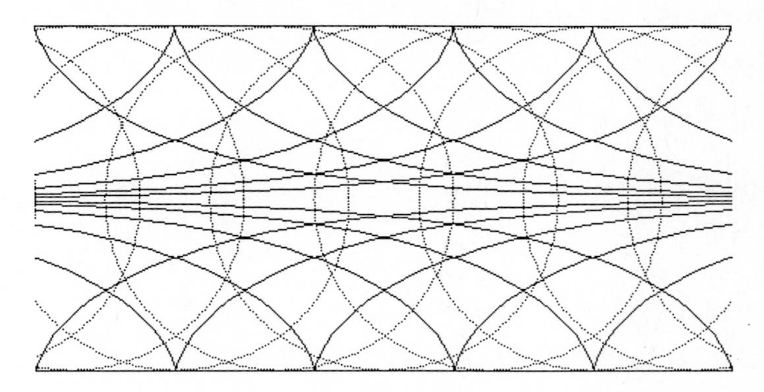

Figure 2

Let $\widetilde{M} \subset S^n \times \mathbb{R}$ denote the lift of the universal cover of $M \subset S^n \times S^1$ to $S^n \times \mathbb{R} = S^n \times S^1$ and set $M = h^{-1}\widetilde{M}$. Then $\hat{M}$ with the metric induced from $\hat{f}|_{\hat{M}}: \hat{M} \to \mathbb{R}^{n+1}$ is isometric to $\widetilde{M}$ with the metric induced from $\widetilde{f}|_{\widetilde{M}}: \widetilde{M} \to \mathbb{R}^{n+1}$. Let $\pi : \hat{M} \to S^n$ denote the projection of $S^n \times \mathbb{R}$ onto the S^n-factor.

Lemma 5: $\pi|_{\hat{M}}: \hat{M} \to S^n$ is a conformal local diffeomorphism.

Proof: By property (ii) at the beginning of this section we have

$$(4) \quad d\hat{f}[T_{(p,t)}(S^n \times \mathbb{R})] = d\widetilde{f}[T_{h(p,t)}(S^n \times \mathbb{R})] = d\widetilde{f}[T_{h(p,t)} S^n \times \{t\}].$$

Together with (ii) of Lemma 4 this implies

$$(5) \quad d\hat{f}_{(p,t)}\left(\frac{\partial}{\partial t}\right) = 0$$

This means that $d\hat{f}_{(p,t)} = d\varphi_t \circ d\pi_{(p,t)}$ where $\varphi_t : S^n \to \mathbb{R}^{n+1}$ is defined as $\varphi_t(p) = \hat{f}(p,t)$. Now as a linear map $d\hat{f}|_{T_{(p,t)}\hat{M}}$ is an isometric embedding into $T_{\hat{f}(p,t)}\mathbb{R}^{n+1}$ and $d\varphi_t$ is conformal by (i) of Lemma 4. Hence $d\pi|_{T_{(p,t)}\hat{M}}$ is nonsingular and conformal. □

Lemma 6: $\pi[(S^n \times \{t\}) \cap \hat{M}]$ is a round $(n-1)$-sphere in S^n for all $t \in \mathbb{R}$.

Proof: For each $t \in \mathbb{R}$ the sphere $S^n \times \{t\}$ is mapped by $\hat{f}$ to a round hypersphere in $\mathbb{R}^{n+1}$. Let $m(t)$ be the center of this hypersphere, $R(t)$ the radius. Then

$$(6) \quad P_t(\hat{f}(p,t)) = 0$$

for all $(p,t) \in S^n \times \mathbb{R}$ where

(7) $\quad P_t(x) = \langle x-m(t), x-m(t)\rangle - R^2(t)$.

As a consequence

(8) $\quad 0 = \frac{\partial}{\partial t} P_t(\hat{f}(p,t)) = \frac{dP_t}{dt}(\hat{f}(p,t)) + dP_t(\hat{f}_*(\frac{\partial}{\partial t}\Big|_{(p,t)})$.

By property (ii) at the beginning of this section the second term on the right hand side of (8) vanishes, so

(9) $\quad \frac{dP_t}{dt}(\hat{f}(p,t)) = 0$.

The zero set of the polynomial $\frac{dP_t}{dt}$ is either a round hypersphere in $\mathbb{R}^{n+1}$ or a hyperplane, depending on wether $R'(t) \neq 0$ or not. Now by the transversality condition in (ii) at the beginning of this section and a compactness argument $(S^n \times \{t\}) \cap \hat{M}$ is a codimension two submanifold of $S^n \times \mathbb{R}$. By (6) and (9) this submanifold is mapped under f diffeomorphically onto a round (n-2)-sphere in $\mathbb{R}^{n+1}$. Then by Lemma 4 (i) $(S^n \times \{t\}) \cap \hat{M}$ is a round hypersphere in $S^n \times \{t\}$, and by Lemma 5 the same holds for $\pi[(S^n \times \{t\}) \cap \hat{M}]$. □

We now exploit our information about the universal cover $\tilde{M} \approx \hat{M}$ to determine the conformal type of M itself. Let $\tilde{p}: S^n \times \mathbb{R} \to S^n \times S^1$ denote the covering projection. The corresponding deck transformations are generated by the map

(10) $\quad \tilde{D} : (p,t) \mapsto (p, t+2\pi)$.

The map $\hat{p} = \tilde{p} \circ h^{-1} : S^n \times \mathbb{R} \to S^n \times S^1$ is also a covering map. Since the vectorfield X^t introduce in the proof of Lemma 4 is

2π-periodic in t, we have $h_{t+2\pi} \circ h_t^{-1} = h_{2\pi} \circ h_o^{-1}$ for all t and the deck transformations for the covering map $\hat{p}$ are generated by $\hat{D} : S^n \times \mathbb{R} \to S^n \times \mathbb{R}$ given by $\hat{D} = h \circ \tilde{D} \circ h^{-1}$, i.e.

$$(11) \qquad \hat{D}(p,t) = (h_{2\pi} \circ h_o^{-1}(p),\ t + 2\pi) \ .$$

By Lemma 4 (i) $\varphi := h_{2\pi} \circ h_o^{-1} : S^n \to S^n$ is a Möbius transformation. We have a commutative diagram

$$(12) \qquad \begin{array}{ccc} S^n \times \mathbb{R} & \xrightarrow{\hat{D}} & S^n \times \mathbb{R} \\ \pi \downarrow & & \downarrow \pi \\ S^n & \xrightarrow[\varphi]{} & S^n \end{array} \ .$$

(12) together with Lemma 5 shows that $M = \hat{M}/_{\hat{D}}$ is conformally equivalent to $\pi(\hat{M})/_{\varphi}$. Moreover Lemma 6 implies that a fundamental domain for the action of φ on $\pi(\hat{M}) \subset S^n$ is the region between the disjoint round hyperspheres $\pi[(S^n \times \{0\}) \cap \hat{M}]$ and $\pi[(S^n \times \{2\pi\}) \cap \hat{M}]$. This means that φ is hyperbolic, i.e. φ has exactly two fixed points. Representing S^n conformally as $\mathbb{R}^n \cup \{\infty\}$ we may assume that these fixed points are 0 and ∞ and φ is given by

$$(13) \qquad x \in \mathbb{R}^n \mapsto \lambda g(x)$$

for some $\lambda \in \mathbb{R}$, $g \in SO(n)$. We record this as

<u>Lemma 7</u>: M is conformally equivalent to $M_{\lambda,g}$.

Using Lemma 7 we will construct for each $\lambda \in \mathbb{R}$, $g \in SO(n)$ satisfying $\lambda \geq \lambda_o$ a channel torus in $\mathbb{R}^{n+1}$ which is conformally equivalent to $M_{\lambda,g}$. This will prove Theorem 4 stated in the introduction.

First note that for each of the two fixed points $p \in S^n$ of g the curve $\gamma : \mathbb{R} \to \mathbb{R}^{n+1}$

$$(14) \qquad \gamma(t) = \hat{f}(p,t)$$

is 2π-periodic, i.e., γ is a closed orthogonal trajectory of the one-parameter family of spheres $t \to \hat{f}(S^n \times \{t\})$ indicated in Figure 3. There are exactly two closed trajectories. Let γ be one of these and let $R(t)$ be the radius of the sphere $\hat{f}(S^n \times \{t\})$. The whole family of spheres (and therefore also the channel torus itself) can be reconstructed using only the 2π-periodic maps $\gamma : \mathbb{R} \to \mathbb{R}^{n+1}$ and $R : \mathbb{R} \to \mathbb{R} \cup \{\infty\}$.

All our constructions so far were independent of the t-coordinate, so we may assume that γ is parametrized by arclength (apply a scaling transformation in $\mathbb{R}^{n+1}$ to make γ of total length 2π). Let N_t be the normal space to γ at $\gamma(t)$. Using parallel translation in the normal bundle of γ we can identify N_t with N_o for $0 \leq t < 2\pi$. For $t = 2\pi$ we obtain in this way the <u>holonomy map</u> $\tau : N_o \to N_{2\pi} = N_o$. τ is an orthogonal linear endomorphism of N_o, which we also call the <u>total torsion</u> of γ. This terminology is justified by the fact that in case $\mathbb{R}^{n+1} = \mathbb{R}^3$ τ is a rotation by an angle α and we have

$$(15) \qquad \alpha \equiv \oint \tilde{\tau}\, ds \mod 2\pi$$

where $\tilde{\tau}$ in (15) denotes the ordinary torsion of γ as a space curve.

Proposition 1: Let $\gamma:[0,2\pi]\to\mathbb{R}^{n+1}$ be a smooth curve, $\hat{g}\in SO(n-1)$ any element that represents the conjugacy class of the total torsion τ of γ. Let $1/R:[0,2\pi]\to\mathbb{R}$ be a periodic function such that the spheres through $\gamma(t)$ which are orthogonal to γ and of radius R envelope a smooth hypersurface in $\mathbb{R}^{n+1}$. Then this envelope is conformally diffeomorphic to $M_{\lambda,\hat{g}}$ with λ given by

$$(15)\qquad \lambda = \int_0^{2\pi} \frac{1}{R(t)}\,dt$$

Proof: By the arguments preceeding Lemma 7 we have to prove that $h_{2\pi}: S^n\to S^n$ is conjugate as a Möbius transformation to the affine map $\lambda g:\mathbb{R}^n\cup\{\infty\}\to\mathbb{R}^n\cup\{\infty\}$. Now this conjugacy class can be read off from the differential of $h_{2\pi}$ at its fixed points. By the arguments at the end of the proof of Lemma 4 the dilatation of this differential can be computed as

$$(16)\qquad [g_{2\pi}|T_pS^n]/[g_0|T_pS^n]$$

where p is the (fixed) point of S^n such that

$$(17)\qquad \hat{f}(p,t) = \gamma(t).$$

Clearly we have

$$(18)\qquad \frac{dg_t|T_pS^n}{dt} = \frac{1}{R(t)}\quad ,$$

which implies (15). The claim concerning τ and $\hat{g}$ follows from the fact that identifying our spheres of radius R(t) via orthogonal trajectories does the same to normal directions of γ as parallel translation of normal vectors. □

<u>Lemma 8</u>: There is a number $K_o(n)$ such that with any prescribed conjugacy class of the total torsion there is a closed curve γ in $\mathbb{R}^{n+1}$ of length 2π whose curvature is bounded by K_o.

<u>Proof</u>: For $n = 2$ (curves in $\mathbb{R}^3$) the Lemma follows easily from the fact that the total torsion of a space curve can be computed as the "algebraic enclosed area" of its tangent image. In higher dimension we build the derived space curves out of pieces which stay in 3-dimensional subspaces of $\mathbb{R}^{n+1}$ which correspond to the invariant 2-planes of the prescribed total torsion τ.

$\square$

Proposition 1 and Lemma 8 provide a method to construct channel tori which are conformally diffeomorphic to a prescribed $M_{\lambda,g}$ if λ is large enough: First construct a closed space curve as in Lemma 8 whose total torsion map is conjugate to g. Then the spheres orthogonal to γ having constant radius $R(t) \equiv \hat{R}$ will envelope a smooth hypersurface if $\hat{R} \geq R_o$ (R_o only depending on K_o). Thus by Proposition 1 any $M_{\lambda,g}$ with $\lambda \geq 2\pi/R_o$ occurs as a comformally flat hypersurface in $\mathbb{R}^{n+1}$.

References

[1] E. Cartan, La deformation des hypersurfaces dans L'espace conforme reel à $n \geq 5$ dimensions, Bull. Soc. Math. France 45 (1917), 57-121.

[2] S.S. Chern, On a conformal invariant of three-dimensional manifolds, in Aspects of Mathematics and its Applications, Elsevier 1986.

[3] M. do Carmo, M. Dajzer and F. Mercuri, Compact conformally flat hypersurfaces, Trans. Amer. Math. Soc. 288 (1985), 189-203.

[4] A. Garsia, The calculation of conformal parameters for some imbedded Riemann surfaces, Pac. J. Math. 10 (1960), 121-165.

[5] A. Garsia, An imbedding of closed Riemann surfaces in Euclidean space, Comm. Math. Helv. 35 (1961), 93-110.

[6] M.W. Hirsch, Differential topology, Springer 1976.

[7] A. Marden, Schottky groups and circles, in Contributions to Analysis, Acad. Press 1974.

[8] H. Reckziegel, Completeness of curvature surfaces of an isometric immersion, J. Diff. Geom. 14 (1979), 7-20.

[9] R. Rüedy, Embeddings of open Riemann surfaces, Comment. Math. Helv. 46 (1971), 214-225.

Aspects of Mathematics

English-language subseries (E)

Vol. E1: G. Hector/U. Hirsch, Introduction to the Geometry of Foliations, Part A

Vol. E2: M. Knebusch/M. Kolster, Wittrings

Vol. E3: G. Hector/U. Hirsch, Introduction to the Geometry of Foliations, Part B

Vol. E4: M. Laska, Elliptic Curves over Number Fields with Prescribed Reduction Type

Vol. E5: P. Stiller, Automorphic Forms and the Picard Number of an Elliptic Surface

Vol. E6: G. Faltings/G. Wüstholz et al., Rational Points
(A Publication of the Max-Planck-Institut für Mathematik, Bonn)

Vol. E7: W. Stoll, Value Distribution Theory for Meromorphic Maps

Vol. E8: W. von Wahl, The Equations of Navier-Stokes and Abstract Parabolic Equations

Vol. E9: A. Howard/P.-M. Wong (Eds.), Contributions to Several Complex Variables

Vol. E10: A. J. Tromba, Seminar on New Results in Nonlinear Partial Differential Equations
(A Publication of the Max-Planck-Institut für Mathematik, Bonn)

Vol. E11: M. Yoshida, Fuchsian Differential Equations
(A Publication of the Max-Planck-Institut für Mathematik, Bonn)

Vol. E12: R. Kulkarni, U. Pinkall (Eds.), Conformal Geometry
(A Publication of the Max-Planck-Institut für Mathematik, Bonn)

Vol. E13: Y. André, G-Functions and Geometry
(A Publication of the Max-Planck-Institut für Mathematik, Bonn)

Vol. E14: U. Cegrell, Capacities in Complex Analysis

Aspekte der Mathematik

Deutschsprachige Unterreihe (D)

Band D1: H. Kraft, Geometrische Methoden in der Invariantentheorie

Band D2: J. Bingener, Lokale Modulräume in der analytischen Geometrie 1

Band D3: J. Bingener, Lokale Modulräume in der analytischen Geometrie 2

Band D4: G. Barthel / F. Hirzebruch / T. Höfer, Geradenkonfigurationen und Algebraische Flächen
(Eine Veröffentlichung des Max-Planck-Instituts für Mathematik, Bonn)

Band D5: H. Stieber, Existenz semiuniverseller Deformationen in der komplexen Analysis

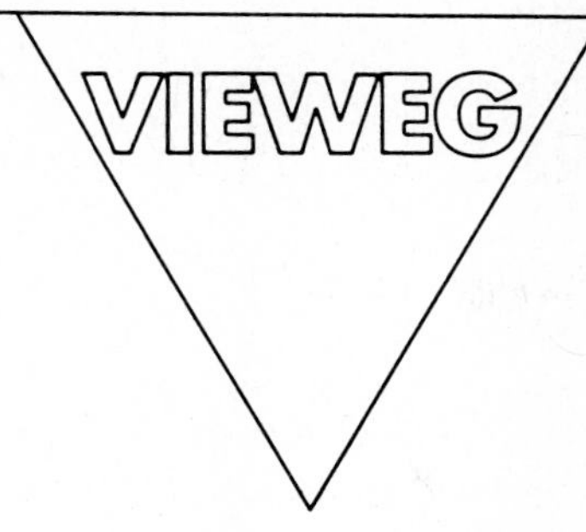

Anthony J. Tromba

Seminar on New Results in Nonlinear Partial Differential Equations

A Publication of the Max-Planck-Institut für Mathematik, Bonn. Adviser: Friedrich Hirzebruch.

1987. VI, 198 pp. 16,2 x 22,9 cm. (Aspects of Mathematics, Vol. E10; ed. by Klas Diederichs.)

Contents: Twisted Tori of Constant Mean Curvature in R^3 *(Henry C. Wente)* – Old and New Doubly Periodic Solutions of the Sinh-Gordon Equation *(Uwe Abresch)* – Global Existence of Small Amplitude Solutions to Nonlinear Klein-Gordon Equations in Four Space-Time Dimensions *(Sergiu Klainerman)* – Global Existence of Solutions of the Yang-Mills Equations in Minkowski Spacetime *(Vincent Moncrief)* – Nonlinear Stability in Fluids and Plasmas *(Jerrold E. Marsden and Tudor Ratiu)* – Lane-Emden Equations and Related Topics in Nonlinear Elliptic and Parabolic Problems *(Wei-Ming Ni)* – Parabolicity and the Liouville Property on Complete Riemannian Manifolds *(Jerry L. Kazdan)* – Global Analysis and Teichmüller Theory *(Anthony J. Tromba)*.

This book consists almost entirely of papers delivered in lectures at the seminar on partial differential equations held at Max-Planck-Institut in the spring of 1984. They give an insigth into important recent research activities. Some further developments are also included.

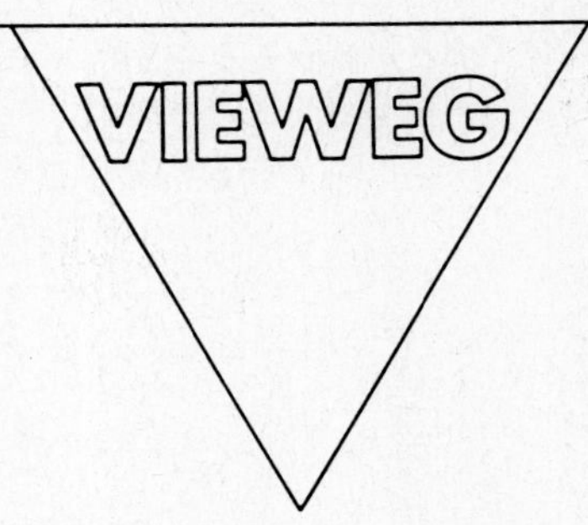

Yves André

G-Functions and Geometry

A Publication of the Max-Planck-Institut für Mathematik, Bonn. Adviser: Friedrich Hirzebruch.

1988. XIII, 231 pp. 16,2 x 22,9 cm. Aspects of Mathematics, Vol. E13; ed. by Klas Diederich.)

The so-called G-functions were introduced as tools into the theory of diophantine approximations by C. L. Siegel in 1929. Their importance for modern arithmetic algebraic geometry became apparent by fundamental research of E. Bombieri (1981). The present book is the first systematic introduction into this field. It is, at the same time, in most parts self-contained and presents many new results. The reader will be introduced into the fundamentals of Fuchsian differential systems, p-adic analysis and important aspects of arithmetic algebraic geometry.